What Will We Tell Our Grandchildren?

WHAT WILL WE TELL OUR GRANDCHILDREN?

The Cultural Origins of America's Climate Inaction

TOM O'NEILL

CLEARSIGHT
BOOKS
Raleigh, North Carolina

ISBN paperback: 978-1-945209-56-7
ISBN ebook: 978-1-945209-57-4
Library of Congress Control Number: 2026913469

Published by Clear Sight Books, Raleigh, North Carolina
Cover design by Howard Grossman

Untitled (Fallen Angel) by Jean Michel Basquiat, 1981 (Photo: Hemis/ Alamy Stock Photo). This work suggests to me the general chaos of our time. Perhaps Basquiat had in mind a much-tarnished version of the American eagle as a symbol of who we are becoming.

Contents

Introduction: What Will We Tell Our Grandchildren? 1

Chapter 1: Overpopulation and Optimism 3

Chapter 2: Malthus and Pessimism 9

Chapter 3: The Lesson from Rwanda 15

Chapter 4: The Threatened Geography of Bangladesh 21

Chapter 5: Foreboding Warnings from Climate Experts 27

Chapter 6: The Rise of Calvinist Puritanism in Europe 37

Chapter 7: Puritanism in the New World 43

Chapter 8: Puritanism Meets the Enlightenment 51

Chapter 9: The Puritan Precedent Set at Mystic Fort 61

Chapter 10: The Emergence of a Secular Puritan Economics 71

Chapter 11: Milton Friedman and the Pursuit of Profit 83

Chapter 12: The Failure of Broadcast Journalism 99

Chapter 13: What Will We Teach Our Children? 115

Acknowledgments 127

Annotated Bibliography 129

Index 145

Introduction

What Will We Tell Our Grandchildren?

SHE WAS ABOUT SEVEN, frolicking in the large whirlpool of the local gym as her father and I talked. I had learned his story in an earlier conversation. For seventeen years, he'd flown some of the most advanced military aircraft ever built. He enjoyed being a pilot, but when back problems began making it difficult to meet the demands of the job, he had decided it was time to retire. Skills he'd acquired could have commanded rich salaries elsewhere, but he'd chosen a desk job in Livermore, California, at one of the local energy labs. The pay was moderate, but between it and his pension, he and his family could live comfortably. Time with his wife and young daughter was important to him.

Our conversation quickly moved beyond pleasantries. Noting the ex-pilot's clarity of mind and lack of bluster, I asked how he'd voted in the 2016 election. He said he hadn't voted for any presidential candidate; neither party had inspired his confidence. I replied, "Surely then you've wanted to step into the breach and promote a politics that offers better candidates." He answered with a dismissive smile—of which I didn't approve—and said such a move was beyond his pay grade.

I gestured toward his daughter. "For her sake, I'd think you'd be concerned." I watched his face. The flash of resentment I'd half expected didn't happen. Quietly, he said he figured things were in sad shape and were going to get worse. He added, "That's why I want to spend time with her and my wife. For myself and them, I want to store good memories."

His sobriety was impressive. But then again, here was a guy who'd spent years keeping his cool at speeds where a split-second mistake could cost his life. I went for broke. Gesturing again toward his daughter, I said, "Just think: If we continue with this vacuum in leadership and continue ignoring climate change and the need for world peace, then by the time she has children her present age, they may think you failed them. Do you think they might resent you?"

I braced for a break in his temper. None came. He replied, "I'm aware of what you're saying. I just don't know what to do about it." We continued relaxing in the warm water once he'd made clear he was resigned to hard times ahead.

I share this anecdote because it strikes me as a conversation in miniature that thoughtful people everywhere may be having with others and with themselves. They know we're in trouble with regard to habitat, but they don't know what to do about it.

This book is an attempt to clarify the challenge of preserving American society and human civilization by exploring the ideas, habits, and histories that have led to our present state of epistemological and political paralysis.

Chapter 1

Overpopulation and Optimism

OPTIMISM CAN BE A PROBLEM. It may seem counterintuitive to make that claim, but think of the many cases in which saying "Everything's fine!" only worsens an already bad situation.

For many, Progress (that is, technological change) is always viewed as inevitable and always assumed to be morally right. This ethos is common to high-tech industries, which tend to regard technical advancement as a benefit even when used for unsavory ends. Nuclear weapons, for instance, may destroy us all, but they are an undeniable example of technological progress.

The care and feeding of these weapons are activities that give my hometown, Livermore, California, its identity. At the Lawrence Livermore National Laboratory (LLNL), the refining of the US nuclear arsenal is well funded and attracts scientists with rare knowledge of nuclear physics. A side benefit for the community is that one meets and moves among very intelligent people—some only peripherally engaged, if at all, in the enhancement of our arsenal. So meeting a former pilot at a local gym wasn't an isolated experience.

While teaching ethics and religion at our local community college, I encountered a brilliant science writer whose job was to write up projects at LLNL and make them somewhat intelligible to nonscientists. Jerry was an autodidact of inexhaustible curiosity. After high school, he abandoned formal education entirely and hired out his understanding to LLNL. Secure and well paid, he decided in middle age to remedy his lack of a college degree, which is how he came to be in my night class.

Our intellectual partnership had an unlikely beginning. A student enrolled in the class brought a friend to audit it. As we approached the end of the first hour, the auditor asked whether, as one who presumed to teach ethics, I'd given thought to how unethical it was of us humans to have captured wolves and intruded on their nature. I responded that I hadn't and, worse, I hadn't given thought to how criminal it had been to take trees and turn them into tables and chairs, and that I hoped during the break the students would take their tables and chairs and place them in the parking lot so the wood could turn back into trees. I was relieved that no one took me seriously.

At the break, Jerry approached me. He said, "You shouldn't have done that."

"Done what?"

"Embarrass that student."

"But he's an imbecile!"

"That's why you shouldn't have done it."

Jerry became my mentor. He furthered my education regarding population growth and its connection to global warming. More than thirty years ago he was already insisting the two were conjoined. The more people, the more carbon footprints, and the more carbon ascending into the atmosphere, the more global warming. He felt the resulting problem was nearly insurmountable but that if people realized what was happening, it could still be solved.

He began with noting the impulse to produce offspring and elaborated on the ways reproduction is constrained by the environment. Constraints come in two major categories. One is physical: limitations of habitat and food supply. The other can be called social: predators, competition with other animal species over turf, male rivalry within a species, and so on. Human success lay in our ingenuity in dealing with these constraints.

The emergence of the human species is not easily dated. It seems to keep getting earlier and depends on what we're willing to acknowledge as evidence of human activity. It seems likely that, in something resembling present form, we came to exist around some fraction of a million years ago; three hundred thousand years ago is an estimate that some are currently comfortable with. Others say we've been around longer, but even then, we'd still be quite a recent emergence in Earth's 4.5-billion-year history.

Much of human experience may have been of the kill-or-be-killed variety. Yet at some point, we learned to make fire. And around twelve thousand years ago—or, if we engage in Carl Sagan's mind game expressing Earth's existence as though it were a 24-hour period, less than a quarter of a second ago—some ancestors learned to configure habitat to institute a recurring food supply and expand its volume. We became aware of seeds and irrigation. Such innovation—complemented by refinements of human consciousness and use of language—was boosted by a strong sex drive and a clear realization that activity as a group enhanced chances for survival. Specialization of skills within a group probably became more commonplace.

We created civilization, which facilitated population growth. In the last four or five thousand years, our rivalry with other species became triumphal. Once we had established settled habitats and a food system providing surpluses, we expanded our civilizations and began engaging in trade among them. We continued to tame the natural world, which we approached with a sense of domination. Our project included domestication of other animals. We cultivated some for their hides. We valued others for their eggs or milk. Some we used as beasts of burden. As we did these things, we incrementally reduced the habitat of those animals for which we found less use.

Success seems to have established in human consciousness a sense of being the masters of our planetary habitat. Likely the leaders among us were optimists. While a notion of "destiny" may have arrived early based on our ability to recruit nature to our service, an explicit articulation of our dominance may be fairly recent. Civilizations in the Mediterranean area and elsewhere were thriving in Homeric times, but it was as recently as the seventeenth and early eighteenth centuries that we entered what we now call the Enlightenment—an age marked by expectations and motivations that endure as backdrop and foundation for the scientific revolution and ongoing industrial and technological age in which we live.

Global population began accelerating rapidly during the Enlightenment, and while it may seem parochial to concentrate on what has been happening to human society in the mere last three hundred years, it's probable that an indeterminate but large fraction—perhaps more than one-tenth of all the humans who have ever lived on Earth—are people who have survived and lived from around the 1700s to the present.

As the Enlightenment dawned, it's conceivable that for the first time a majority of people began to take for granted that human success would persist and that the lives of humans would continually improve for as long as the planet itself endured. The apocalyptic residue from religious rivalries in the sixteenth and early seventeenth centuries notwithstanding, those who were prominent in the Enlightenment weren't inclined to envision a collapse or some terminal completion of the human project in their foreseeable future.

One finds optimism in the writings of the economist Adam Smith in Scotland. Edmund Burke, the English parliamentarian from Ireland, shared it. So did Alexander Hamilton in America. Optimism seemed to become a generally held disposition of intellectuals throughout Western civilization, even as continuing political turmoil, such as the Reign of Terror in France, could have suggested a future that was less promising. In spite of such turbulence and poverty, confidence became the fashion.

I belabor this because, as I hinted earlier, optimism can become a bad habit. In the opening decades of the nineteenth century, the philosopher G. W. F. Hegel gave our growing confidence perhaps its most thoroughgoing exposition. He argued that our expectation of continuing upward development was in fact rooted in our intimation of the unfolding life of the One Being of which we, and all phenomena, are a part.

Today, Hegel is probably most remembered for what was later codified as dialectical structure (thesis–antithesis–synthesis) as an account of how ideological development occurs. Less known is how this structure was, for Hegel, a subdivision in the life of Absolute Spirit as It makes demands on Itself in a struggle toward final realization. Spirit creates, then opposes what It has created, then reconciles those two moments in a third moment of fresh creation—and so on. This meant that destruction itself was creative. In Hegel's formulation, even war shouldn't demoralize us—it is a needed ingredient for cultural development.

I mention this to indicate how completely an attitude of optimism became installed in many nineteenth-century Western intellectuals. For the majority of humans, less prone to skylarking, it may have been possible, if they listened to Hegel, to reduce his message to something more familiar. One might say that his theory implied that human rationality was now summoning all things into the service of Progress and was making a positive use of everything, including war. Even skeptics among inhabitants of the nineteenth century may have felt complacent in the belief that our species would go on improving its status forever. Whether God was or was not, surely humans were destined to progress.

But here, in Jerry's account, he called up a preemptive counterthrust.

Chapter 2

Malthus and Pessimism

If OPTIMISM WAS THE THESIS, what was the antithesis that predicted the bad feeling now common throughout the world? In the closing decade of the eighteenth century, a nay-saying shadow-caster arrived. Thomas Robert Malthus seemed to have come to the party to spoil it. A well-established clergyman in England's intellectual class, Malthus sounded an alarm in the sunny evening of the Enlightenment. Recognized as a gifted mathematician, he claimed as a discovery of mathematics that while the increase in food production occurs arithmetically, the increase in population occurs exponentially. Impressed by the rapidity with which the population of the American colonies had grown, he claimed that the population of England would soon outstrip its supply of food unless the English imposed restraints on their tendency to breed.

Malthus thought population tended to increase much as compound interest does. In the first decade of this century, American homebuyers learned that interest generously suspended for the moment gets folded with unkind rapidity into principal. The rapidity was such that mortgage payments ballooned out of reach for an estimated 3.8 million Americans who eventually lost their homes to foreclosure. At the end of the eighteenth century, Malthus predicted that an unforeseen foreclosure dealt by nature itself was awaiting Britons if they continued to have large families.

It is the tendency of our species, he thought, to succeed into failure. While he can be second-guessed on timing, it is his warning that population could eventually outstrip food supply on which his title to genius stands or falls. The optimism people feel as they look at land beckoning from the horizon inspires them, he thought, to have larger families than that new acreage can support. People's tendency then is to seek virginal land yet farther away that can feed the new surplus of humans. This may sound fine, but since his studies indicated that each year births were exceeding deaths, Malthus predicted that Britain's agricultural land would prove inadequate. Soon the population would outgrow the foodstuffs the land could supply, meaning the country would no longer be able to sustain its people.

Why hadn't this discrepancy become apparent earlier than the end of the eighteenth century? Malthus may have surmised that since tilled acreage and population were both smaller in earlier times, the discrepancy between their rates of increase may not have been readily observable. In his day, however, the discrepancy had become apparent—at least to him.

A common insinuation against Malthus is that he had no sympathy for the poor. This is unfair. What he hated was that anyone *was* poor. It is true Malthus didn't like the English Poor Laws, believing they institutionalized poverty. He was likely aware as well that the plight of the poor can register as an asset in the account books of the rich. As any well-endowed landowner or industrialist knows, when there are many destitute people, the cost of labor is low.

History, as we practice it, has been chiefly a history of the well-to-do by the well-to-do. Perhaps Malthus thought that we had overlooked the discrepancy by which population had exceeded resources because the burden of inadequate food falls first on the already disadvantaged. Children tend to be its first victims—something confirmed routinely in many parts of the world today. Malthus may have thought that so long as the comfortable classes of England could have their afternoon tea, recordkeepers and historians would tend not to notice the plight of the poor. (About a half century later Charles Dickens, though antagonistic to Malthus, would make a recurrent theme of his novels the charge that the well-to-do were indifferent to the poor. This was a prominent theme in *Oliver Twist*, for instance.) In wishing to limit British population, it was Malthus's wish to see brought into the world only as many people as Britain could decently provide for.

He first published his observations on population growth in 1798. Today—more than two centuries later—saying that Malthus was wrong seems to have become an academic fashion. There is a difference, though, between being wrong and simply miscalculating. The fault in Malthus's calculations was in not allowing for the immense amount of arable land still available to the British in other parts of the world—having as they did a Royal Navy that enabled them to take control of many lands abroad. Malthus didn't foresee that colonialism would enable the British population to continue to expand smoothly for well more than a century after its demands had outstripped domestic resources.

Another thing Malthus didn't anticipate was the coming boom in cheap energy. There would be large-scale technological innovations in farming that would augment the wealth the landowning and middle classes of England would accrue by exploiting labor and land in what came to be called the Third World. (Though it is now dated and often put to objectionable use, I find the phrase "Third World" still useful—and preferable to the phrase "developing world" with its gratuitous implication that things in general are improving globally.)

While Malthus himself was personable and gracious, his predictions registered as unrealistically gloomy at the time they were made. It wasn't long before *neo-Malthusian* became a dismissive term.

The main thing that preserves the relevance of Malthus is something he didn't know about: the role that human contributions of carbon dioxide into the atmosphere would play in corrupting the planet as a habitat for humans.

A substantial amount of what I set forth here echoes Jerry's take, which became generally mine. While the timetable Malthus presented proved too short, the insight he promoted was sound. And while he wrote mainly about the prospects of the British, his central conviction was that resources available to humans are finite even though our appetites and goals are open-ended.

We humans have no envisioned limits—either to the increase of our numbers or to the power of the nationalisms from which we seek identity. Malthus's apprehensions about the open-ended ambitions of most societies we form seem currently validated. (Though what happens in American civilization, I'll contend, may owe more to a Calvinist quest for the "signs of election" than the non-Calvinist Malthus realized.) What Malthus establishes for those who listen is that there's a conflict between the growth of population and the maintenance of habitat. We can see validation of Malthusian thinking in the case of America. A desire to quickly open the vast territory occupied by Indigenous Americans beyond the Appalachian Mountains was one motivation for our Revolutionary War. And from at least the days of Andrew Jackson, we Americans had decided to gobble up the West as fast as we could.

We had barely given young people the advice to "Go West!" when we heard with dismay about "the closing of the American frontier." Frederick Jackson Turner informed us in 1893 that the West—imagined by settlers as a place of near-infinite, untapped opportunity—was occupied. A generation later, the Dust Bowl confirmed the general bad news about land use when the American West faced the intersection of the loss of frontier and a desiccation of farmland, leading to severe hunger.

Before that, the ghost of Malthus had been whispering in the ears of some politicians—men who thought about land management. They did not want a disorderly free-for-all to substitute for well-managed development. John Quincy Adams, our sixth president, and his secretary of state, Henry Clay, were agents serving this concern. The name of their proposal was "the American System," and they wished to assign the federal government a firm hand in guiding an incremental development of the West. One may regard them as post-Hamiltonian Federalists or as pre-Keynesian Keynesians. Andrew Jackson campaigned in 1828 on a promise to throw out their federal approach to banking and infrastructure, and once elected, he did as promised.

One can sense a kindred recklessness a half century later in the honor paid to George Armstrong Custer, who in 1876 led his Seventh Cavalry in an effort to kill thousands of Sioux—men, women, and children—gathered at the Little Bighorn River. Custer failed in his attempt, but American citizens, for a long time after, celebrated his intentions as a heroic effort to secure lands for a growing population of White Americans.

Steve Jobs, Mark Zuckerberg, Jeff Bezos, Rex Tillerson, Elon Musk, the Koch brothers, and US war departments—each in their own arena, and with many a small-scale imitator—seem to be expanding our list of human "needs" so rapidly that it dizzily churns forth new demands, giving fresh credibility to Malthus's concern that a growing population reduces habitat—for more and more habitat must be recruited to serve all these needs.

Even now, when riotous weather and brutal pandemics should instruct us, Americans cling to the American Dream, in which we expect each generation to live more extravagantly than the preceding one. High aspirations are understandable among members of minorities long excluded from extravagance, or for recent immigrants—people whose share in American affluence is often meager for a time. I'd posit though that no mere redistribution of goods and services will render the current American lifestyle sustainable.

We spin into vertigo over the variety and multiplicity of fresh projects and products. We're at a loss to see how to fit them into a harmonious pattern—our tolerance for pollution doesn't incline us to do so.

In an inaccurate reading of Adam Smith, our gurus and executives tell us a so-called "invisible hand" will sort things out. They seem to feel there is no such thing as extravagance, and thus they continue to expand their projects, seemingly without a clue as to the long-term destruction that is their cost.

Chapter 3

The Lesson from Rwanda

"IN 1992 [RWANDA] HAD the highest growth rate in the world, an average of 8 children for every woman." This sentence captures the point at the center of an analysis in *Consilience* by Harvard biologist E. O. Wilson, in which he traces overpopulation to a Malthusian tendency to succeed into failure. Wilson points toward a mechanics of overpopulation commonly in operation and claims the tragic case of Rwanda illustrates this process. He suggests that when things go wrong there is an increase in the birth rate that makes trouble—or at least that a high birth rate intensifies the existing problem. Rwanda, he argues, serves as a harbinger of unwholesome things to come in an increasingly desperate, heavily populated world generating climate change.

During the 1990s, Rwanda saw a long tribal rivalry evolve into a project of genocide initiated by radical leaders among the majority tribe. It was unexpected, because even though Rwanda's past had been marred by a Hutu massacre of Tutsi at the start of the 1960s, the history of rivalry between the two ethnic groups had usually been tempered by civility. The Tutsi minority had exercised rule, and the six-to-seven-times-larger Hutu majority had accepted a kind of informal subordination. It had been a class system, but a familiar one—one with the benefit of predictability.

In the first half of the twentieth century, first the Germans and then the Belgians intervened in Rwanda as colonial supervisors. Claims were no doubt made that this was done from noble intentions: The West was going to bring the benefits of civilization to those who lacked them. This benign profession notwithstanding, we can be confident that the question actually in the forefront of colonizers' thinking was: "How can we best utilize the resources and people of Rwanda to our advantage?"

The answering European agenda fit the colonial pattern: "Let's recruit what seems the more dominant faction and thereby achieve leverage over the whole." Thus, the intensity of the ethnic rivalry was heightened. Rwanda became a pawn in European competition and was exploited by those powers.

It's important to notice though—since it can easily get overlooked—that the presence of Europeans accomplished some things in Rwanda that most Rwandans regarded as improvements. Ironically, it is to these improvements that Wilson, like a modern Malthus, turned to seek the springs of genocide. Modern medicine was to Wilson as fresh agricultural land was to Malthus: It led to dramatic increases in population.

In the period from the European arrival in the early 1900s to the genocidal outbreak near the end of the century, infant mortality in Rwanda declined greatly. One demographer estimates that from 1930 to 1990 the population increased about 450 percent. What may have most enabled this extraordinary growth was hands-on administration by the Belgians, which began in 1916 during the First World War. Sanitation improved.

Contagious diseases were reduced (though even today they remain a major problem). Due to agricultural reforms and foreign trade, the diet of Rwandans became more varied and plentiful. As mentioned, and of great importance, modern medical practices led to a far higher percentage of newborns surviving into childhood. Modern medicine also led to increased longevity among adults. As the number of adults of reproductive age rose, Malthus's predicted exponential rate of population growth became visible.

Rwandan society had long endorsed frequent pregnancies, knowing that a high infant mortality rate meant a woman could bear many babies and still have a small family. As late as 1930, life expectancy was estimated at less than forty-four years, so a woman could not realistically expect a long span in which to nurture her surviving children into adulthood. Consequently, her culture encouraged her—alongside her own tendencies, one would think—to begin bearing children at an early age.

The factors listed here changed Rwanda dramatically, especially from the 1930s forward. Occurring incrementally though steadily, this transformation may not have been recognized as a phenomenon of great importance, much less as something foreshadowing genocide. And over decades, there was no radical transformation of culture. As a society that before the 1994 genocide was heavily Roman Catholic (about 40 percent of the population in the 1960s, peaking around 65 percent in 1990; Protestantism accounted for about 5 to 15 percent over the same period), Rwandans continued to favor frequent pregnancies during a woman's childbearing years. This quest for children, of course, wasn't mysterious. No doubt it was motivated by the universal joy the birth of a newborn brings, by the hope that "many hands would make light work," and probably by a desire to have adult children to care for parents in their declining years.

The consequence of this mix of factors was a "baby boom" in a small country where the development of food and water resources encouraged population growth but was not keeping pace with it. While the genocidal intent that emerged is usually assigned to the Hutu (with the Tutsi as victims), the truth may have been that, at least at a subconscious level and well before the 1990s, both ethnic groups had begun to regard the increase of the other group's offspring with misgiving. In the matter of reproduction, consciously or unconsciously, I suspect that something comparable to an "arms race" was brewing.

I arrive at this hypothesis from the need to explain what is almost inexplicable. A mature Hutu would look uneasily at the increase of Tutsi grandchildren; a mature Tutsi would look uneasily at the increase of Hutu grandchildren. It's plausible that imaginations in both groups gradually became preoccupied by a specter of scarcity: "Will there be enough for our grandchildren if the needs of those other grandchildren are met?" Tribal consciousness came to the fore. In the last third of the twentieth century, the concept of "otherness" became potent in Rwanda, three-dimensionally present. The rational thing might have been to initiate nationwide conferences regarding the need to curb the climbing birth rate. But Malthus has seldom been a welcome advisor in human affairs.

Instead, there were rumors and stereotyping and maybe false-flag activities—moving in a quickening tempo into violence and retaliation by the Hutu in 1959. Then, three years later, there was an inexplicable moment just before the Belgians withdrew from Rwanda. The Belgians, who had long favored the dominant Tutsi minority, suddenly shifted their support to the Hutu majority. When they withdrew, it was as though they left a loaded gun behind. Decades of turmoil and refugeeism followed. By the early 1990s, bolstered by confidence in their numbers, some hardline Hutu decided to pursue a "final solution."

Statistics become weapons in time of war, and it is hard to know just how many Tutsi died. Some claim that more than half of all Tutsi in Rwanda were killed; others go so far as to say three-fourths. On the other hand, there are scholars who claim to document a reciprocal effort by some Tutsi to eliminate all Hutu. If they're correct, it means two parallel genocidal programs were being implemented simultaneously. In the case of Tutsi, however, their small numbers and their later conduct toward the population as a whole put in question for me whether a significant number of Tutsi were genocidal. Traditionally, their program hadn't been ethnic cleansing; it had been to be in charge.

When the conflict subsided, one estimate was that about 10 percent of the Rwandan population had perished in civil war. The deaths reduced resource pressure, though at an unimaginable cost in human misery. Every survivor knew many people who had perished. And because—for all Malthus's warning—human beings respond to misery by attempting to increase their numbers, it shouldn't surprise anyone that in the nearly three decades since the genocidal war ended, the number of Rwandans has again grown exponentially. Some demographers project that by 2050 the population will be more than double what it was in opening teens of this century.

E. O. Wilson said, "Rwanda is a microcosm of the world." He seemed to be inviting us to consider what it means if Rwanda's turmoil proves to be a paradigm for predicting genocide in other regions as habitat succumbs to global warming caused by a reckless use of resources. Even though many parties may concur that Malthus was wrong, the way they behave—in lust for other peoples' land—endorses Malthus's concerns.

Chapter 4

The Threatened Geography of Bangladesh

NOW LET'S LOOK TO BANGLADESH—for its own sake but also for clues to times to come in other places.

I view Bangladesh very much from the stance of an outsider. When it comes to politics, it appears to me that for four thousand years the inhabitants have been trying to establish political stability but that some external power is almost always trying to recruit the geography and population for its own agenda. At the same time, there have almost always been local leaders trying to institute domestic improvements for their communities. Some of this has been going on right into the twenty-first century.

In 1947, when Britain ceded independence to India, Gandhi opposed Partition and wanted what would eventually become Pakistan to be part of a unified, liberated India. Despite his efforts, Partition went ahead, and Pakistan was formed with two wings, West Pakistan (modern Pakistan) and East Pakistan (modern Bangladesh). This was an awkward arrangement, because a large section of India separated East Pakistan from West Pakistan. Also, the people of East Pakistan felt marginalized by the West Pakistani leaders. Furthermore, some East Pakistanis felt the West

Pakistanis had attempted a genocide against them, while many West Pakistanis denied any attempted genocide had taken place. After about two and a half uneasy decades, in 1971 East Pakistan declared independence, triggering a war with West Pakistan. With India's support, East Pakistan won independence, becoming Bangladesh. Turbulence, however, continued. Bangladesh has experienced cycles of parliamentary democracy, corruption, military coups, suspensions of democratic rule, and restorations of democratic rule.

My subjective sense is that the Bangladeshi are a valiant people who won't give up on themselves. Economically, they've done well. They've established themselves as leaders in the international garment industry. They've maintained a strong, cordial trade relationship with China. They've incorporated strengths of the British education system into their own schools. They've recognized and tried to root out corruption in their bureaucracies.

At this point in time, Bangladesh faces many challenges with respect to water—including both access and quality—that are exacerbated by global warming. Adding urgency is the fact that Bangladesh is a densely populated country. The infant mortality rate is moderately high (especially relative to wealthy countries), which might lead one to predict a coming decline in population. A possible explanation for why no decline has been seen is that, as in the case of Rwanda, this frequency of infant death functions as motivation for fertile women to have many children. The population in Bangladesh is, in fact, growing rapidly. There are approximately three times as many live births as there are adult deaths in Bangladesh each year.

For the year 2050—a quarter of a century from the time at which I write—the French Institute for Demographic Studies

(INED) projects a population increase of 39 million people beyond the 176 million it records for 2025. Only nine nations are projected to have a larger population. It seems unlikely, though, that Bangladesh will reach the number projected for it. While INED's prediction for 2050 is not as high as some earlier predictions, my hunch is that the projection remains a demographer's extrapolation more than a climatologist's, agronomist's, or sociologist's prediction.

Summer arrives sooner and reaches warmer temperatures than in the past due to the more rapid springtime melting of Himalayan glaciers to the north. At least two problems develop. As glaciers melt at a quicker tempo, there is less time in which to utilize the rapidly flowing water. I'd be surprised if the demographers have factored this in. Further, due to faster evaporation, there is less water to work with. This new pattern disrupts the traditional agriculture of Bangladesh in the north and central regions.

As for the south, again there are water problems developing. The rapid water of the fast-flowing rivers carries debris to the southerly delta region, dislodging fertile soil as it travels and releasing some of that soil into the Bay of Bengal. That's bad enough, but in a sense, the bay fights back. As its tides reach higher, they rinse the fingers of the delta with salt water—making them less apt for agriculture.

Adding to the strain on food supply, waves of Rohingya refugees—totaling upward of seven hundred thousand since 2017—have crossed into southern Bangladesh from Myanmar, fleeing ethnic and religious persecution. (Perhaps this is in response to an effort by the government to free up oil-rich land in Myanmar.)

Rohingya refugees crossing the Naf River from Myanmar into Bangladesh (Photo: Moises Saman/Magnum Photos). We—along with others—are agents of the refugeeism that is now happening at an increasing tempo.

We are free, of course, to blame the Bangladeshis themselves for these problems. It doesn't seem biased or unkind to observe that since the end of British rule in 1947, Bangladesh has been a place of near-constant political turmoil. The British may claim, of course, to have been impeccable overlords; however, what seems to have happened after they left was an unpredictability in rival bureaucracies. Additionally, as population has increased, arable land has dwindled. Climate change has shortened the growing season, and people are flocking to the cities—especially to Dhaka, where there are jobs but low-paying ones.

The people of Bangladesh maintain charm in the face of adversity, but in the near future, their faces may shrivel with hunger. Search out the photographs and videos. The people there greet the camera with a smile, perhaps hoping that the visitors who witness their plight won't just spend money among them but, more importantly, will report the story of their growing desperation and summon a response from outsiders. Their trust reaches out to us. They have minimal responsibility for the climate change that is increasing the difficulty of their lives. (Both their per-person and total emissions are below the global average and *far* below those of major emitters like China, the US, and India.)

Suppose we Americans, who have major responsibility for climate change, blink and look elsewhere. If we deflect our attention (something we now do in the face of refugees), it's hard to see how we can have self-respect or deserve respect from others. If we will not extend to others the help we ourselves would hope to receive in such circumstances, what claims can we make of moral standing?

To sum up current reality: As we Westerners warm this unique planet (so far, the only place in the universe we can be sure provides habitat for our form of life and consciousness), we watch the Ganges and Brahmaputra Rivers—both traditionally friendly to human purpose—now flow earlier and faster than in the past. Summer months arrive sooner and are warmer. And while this goes on, an incremental rise in ocean level is starting to salt the delta in the south. We conclude: This densely populated country (where people seek a remedy to misery by a tempo of births that heightens misery), in which the arable land both in north and south is contracting, is on its way to becoming desperate.

One senses that hardly any planning is currently going on to prevent predictable calamities. Such calamities, if left unaddressed, will likely register as catastrophic by 2050. In our best-outcome-for-Americans scenarios, most people in America who are now forty years old or younger will live to see increasingly turbulent episodes of dislocation and refugeeism arising outside our borders. If we don't wish most Americans to inhabit some weird combination of Fortress America and Mad Max World before they die, we must begin now to create the alliances and institutions and technological innovations that can reduce chaos. Water conservation will be an important ingredient of any mitigation that occurs.

Chapter 5

Foreboding Warnings from Climate Experts

AMONG US ARE CONSCIENTIOUS SCIENTISTS and their disciples who are trying to make a noise about climate change. Here's a daunting paragraph from Joel K. Bourne, Jr., agronomist and former senior editor with *National Geographic*:

> According to the IPCC [Intergovernmental Panel on Climate Change], the impact of climate change on the world's freshwater supplies will likely vary widely from region to region, with arid and semi-arid regions getting drier and other areas experiencing heavier, more variable rainfall and more frequent flooding. They predict the negative impacts will outweigh any benefits.

As we approached Christmas 2017, semi-arid California experienced wildfires in many places, especially just north of San Francisco Bay and then not far to the east of it. This occurred at a time when the rest of the country was moving into nearly unprecedented storms, flooding, and chill. Demonstrating our capacity for selective vision, most who reported on these phenomena described only "intense weather" and "natural disaster" and made no mention of "global warming" or "climate change" nor gave any suggestion that we humans might be the cause of what we were suffering. Much precious water was poured onto the fires; eventually, when stormwater descended, the water-sodden California topsoil flowed laden with debris into sewers and then went on to contaminate the ocean. In autumn 2020, we experienced an even harsher recurrence of such phenomena. (What to say about such things? NBC News has risked offending advertisers by speaking of climate change. By contrast, ABC seems to have maintained terminal phases of cowardice. CBS, of which I despaired once Scott Pelley was exiled from the evening news, may be somewhere between NBC and ABC. A loss of integrity is probably not due to the anchors or the staffs that serve them but, as will be discussed at length in a later chapter, to the executives, whose interest in the news seems largely shaped by its function as bait to get people to watch unsolicited ads.)

Bourne followed the prescient paragraph with which I opened this chapter with an accumulation of daunting predictions:

> Less snowfall and depleted aquifers present the farmers of the future with a huge challenge. A recent study by the FAO [the United Nation's Food and Agricultural Organization] estimated that our increasing population and incomes around the world will require an 11 percent increase in irrigation withdrawal by midcentury. If you add the hotter temperatures of climate change to the mix, the problem starts to look like the curse of the Danaides, the 49 brides of Greek mythology who slew their grooms and spent eternity hauling water in a fruitless attempt to fill a basin full of holes. The math

is simple: a 1°C rise in temperature results in roughly a 3–4 percent increase in water demand. The FAO predicts that climate change will cause global irrigation demand to increase by as much as 20 percent.

One can see where this is going. As there are more and more people, and as more of these people expect a higher standard of living, there will be more drinking of water, taking of showers, flushing of toilets, raising of thirsty livestock, and eating of food that demands irrigation—all at just the time when much arable land will require increased irrigation due to global warming and the resulting aridity.

Regarding fresh water, Bourne continues:

> A more detailed study by the USDA [United States Department of Agriculture] in 2008 projected a 35–64 percent increase in demand in the United States by just 2030. In perhaps the most disturbing estimate, after spending five years putting together a *Comprehensive Assessment of Water Management in Agriculture*, 700 of the world's top water experts concluded that, without significant gains in productivity in the world's rain-fed fields, water demand by agriculture will nearly double by 2050—a quixotic vision, since the water for such an expansion just doesn't exist.

A number of responses to this bleak statement are possible. One can say, for instance, a new Green Revolution is now due in the form of more wonderseeds. However, Norman Borlaug, the pioneer of the original Green Revolution, was notably skeptical about any such breakthrough. One can say, "Well, he and the seven hundred other 'world experts on water' are a bunch of skylarking publicity hounds." A rebuttal to such an assessment might be that celebrity-hunting hounds are not noted for engaging in long-range planning (nor are water researchers generally known for being preoccupied by celebrity).

A quieter response, more attentive to reality and not engaged in name-calling, might be: "Looks like there's going to be a significant die-off of humans before this century is over." That hunch has high plausibility. Such a politically impolite statement is bolstered by evidence that a die-off already seems underway in parts of the world.

Bourne's own solution of choice looks to conservation—both through general lifestyle change and through more precise irrigation technology—grounded in a realistic understanding of the preciousness of water. A steep decline in the rate of human reproduction would help too, according to Bourne.

Bourne doesn't seem to entertain hope for a remedy by way of desalination. He may feel the high cost makes it a nonstarter on economic grounds—though one might think the urgency of the water predicament would override reflections on cost-effectiveness. James Hansen, director of Climate Science, Awareness and Solutions at Columbia University's Columbia Climate School, has spoken of desalination with some interest. Great as the distaste for nuclear power may be as a result of incidents at Three Mile Island, Chernobyl, and Fukushima, these costly and alarming events might best be counted as learning experiences. There is reason to believe that new designs can produce nuclear power plants safe from everything but terrorists, meteors, earthquakes, tidal waves, and volcanoes—and even some of these threats can be reduced by newly designed shutdown systems.

Hundreds of new plants dedicated to desalinating seawater could conceivably become a major option for mitigation of water scarcity. Demands of thirst and hunger cannot easily be dismissed. When others wave off nuclear power with a flip of the wrist, I question whether they've taken the full measure of our predicament.

It haunts a thoughtful mind that attention to predictable aridity and thirst is exercised by only a few. Most people are marking time. While they wait for someone to do something, we are in fact doing a great deal—we're maintaining and expanding a lifestyle that will destroy many of what we regard as life's amenities long before our grandchildren reach old age.

In 1831, Alexander von Humboldt, a scientist acutely attentive to the web of relationships constituting habitat, speculated that the increasing release of gases into the atmosphere and the maintenance of gases there would eventually affect human habitat negatively. But his was not the first attention to this subject. In previous decades of the 1800s, Joseph Fourier had been interested in how heat is stored in the atmosphere. Fourier died in 1830, but subsequently Claude Pouillet, whose work focused mostly on electricity, also became interested in heat storage. He credited Fourier as a thoughtful pioneer in the study of heat, and he followed Fourier, quietly studying interactions of temperature and gases. The metaphor of the "greenhouse" may not have become familiar until later, but some think he was the first to use it.

Or perhaps the first to use the metaphor was Svante Arrhenius in 1896, about thirty-seven years after the death of von Humboldt and twenty-eight after the death of Pouillet. It may be that it was only with Arrhenius's work that scientific efforts to quantify the correlation of gases and temperature migrated from a few labs to become a matter of general interest. Arrhenius, a Swede, was himself optimistic about the prospect of global warming, because he thought it would provide insurance against any return of ice ages and would open new northern climes to farming.

It wasn't for another half century, though, with the industrial boom that followed the Second World War, that a new take on the matter would form. In the International Geophysical Year (IGY) initiative of 1957–1958, about thirteen years after the end of the war, the Keeling Curve became an item of discussion among climatologists. It was a graph of carefully collected data that, as more yearly data were collected, indicated a notably faster accumulation of greenhouse gases in the atmosphere than had been generally expected. Impressively, this alarm didn't come from a hypothesis to explain some current inconvenience but was generated from a forecast of inconvenience not yet much manifested. It was like identifying the source of an epidemic before it became an active threat. The patterns are similar. The urgency coming into climate science was, of course, grounded in experience, but one can say it was less a *response* to phenomena at hand than it was a *prediction* of phenomena to come. That the prediction has proved true is, a rational person must concede, a powerful indication that the scientific methodology involved is wholesome. That so many of the citizens of the United States have refused to bring this insight into the conduct of their daily lives is a strong indication that we are probably not going to surmount the crisis that challenges us. Instead, we have sought out a leader who has given us permission to worry not at all about the challenge in front of us.

Thoughtful people today don't speak of "Newton's theory of gravity"; we take gravity as a given. We can still elaborate on it, refine our understanding of it, restate it, and supplement it with new discoveries and hypotheses, but because gravity's presence and some of its consequences are confirmed science, we don't speak of a theory but of a *law*.

The objects discussed in climatology may strike us as more complex and obscure than the law of gravity, but these objects are no more conjectural than gravity is. The warming is a given. Can the description be refined? It can be and is, in fact, being refined all the time. A preoccupying activity of worldwide climatology over the last half century has been in refining data on global warming and how quickly it has been occurring.

This doesn't mean everyone accepts such confirmations. Just as there are groups in the US who preferred to say "What epidemic?" and then let adults and children die of COVID-19, so too are there swaths of the population who prefer to say "What global warming? The weather's quite normal!" They say this as they watch their homes being swept downriver.

So the blanket of gases grows thicker, and the amount of heat being trapped increases. The effect has an exponential character due to negative synergy: As ice melts from the heat, there is less shiny surface on the earth to reflect light and its warming effects back into deep space. The faster that ice melts, the faster yet it is going to melt. Nonetheless, we regret the high cost of gasoline and the interruptions of air travel that storms cause—not recognizing the real expenses involved.

Planetary heating so far hovers a little over one degree Celsius above preindustrial levels (mid-1800s). Scientists warn that 1.5°C of warming is a critical threshold. Beyond it, the risks of extreme heat, sea-level rise, and ecosystem collapse increase sharply. According to the World Meteorological Organization, the years 2023 through 2025 have been the warmest on record, with a consolidated three-year average increase of 1.48°C, with individual months exceeding the 1.5°C mark. Climate researchers such as James Hansen have cautioned that once we cross certain tipping points—for example, the loss of Arctic summer ice or the destabilization of major ice sheets—it's difficult to reverse changes. And if we remain too long in the danger zone above 1.5°C, return can become impossible.

Refusing to take in what our eyes disclose, many of us assume that the climate will remain pretty much as it has been in recent memory. We refuse to accommodate. But the traditional life cycles of many plant and animal species are already being disrupted, and if the heating reaches a point that dooms the ice at one or both poles, we will forever be "not in Kansas anymore." Close observers of our species's behavior declare this is almost certainly going to happen—probably within the lifetime of today's younger generations. This is why serious people take seriously books with titles like *Losing Earth* and *The Uninhabitable Earth.*

Is this crisis genuine? Yes. As the amount of trapped heat increases, the Earth's capacity—starting with our forests and oceans—to absorb new heat, and thereby spare existing habitats from drastic harm, is being exhausted. Perhaps we should linger here. We need to register this. Something more drastic than mere change in habitat is being predicted—and is already reported from many different locations. A translation into lay language of what climatologists are reporting, often in quiet tones to shield the public from despair, is: "The planet we know is going out of business." Those who currently ridicule advice to change dietary habits and transportation customs will, if they prevail, bring a far more radical imposition of behavior revision upon us than any of us want.

The Miami shoreline (Photo: steheap/adobestock.com). The ground on which these skyscrapers stand is sandstone, limestone, or a combination. Add to this vulnerability the regularity of being battered by hurricanes, and then add what even a slight rise in ocean level can do—a cause for concern.

One might say human habitat dwindles twice over as world population grows. It dwindles once from physical forces, and dwindles again from social behavior. If you order pizza for six and six additional guests turn up, each guest's share—supposing all guests are treated equally—is reduced by half. Supposing then that the pizza order was already in trouble when you placed it (the ingredients being already scarce). It's not hard to see that your order is in trouble from two directions. It is this way with climate.

Bringing one to the brink of vertigo and despair is the realization that the negative consequences are operating exponentially. Functioning as compound interest does, with past harm continually folded back into the base for increasing quantities of harm to come, the complex system—synergistic and exponential—predicts the arrival of a catastrophe beyond anything Malthus could have imagined.

Chapter 6

The Rise of Calvinist Puritanism in Europe

IN THE 1500s, MARTIN LUTHER and the Reformation became a countervailing force to the Renaissance then radiating westward and northward from Italy. In part, the Reformation was a protest against the ongoing appropriation of pagan Greek and Roman culture by the reigning Italian culture, Roman Catholicism. Luther and other northern reformers felt that in this Latin merging of pagan and medieval cultures a contamination of faith was taking place, and that religious concern for the world to come was yielding ground to a spirit of living for the opportunities and pleasures of the world at hand.

An abundance of evidence backs up the charge of emergent worldliness—for instance, the corrupt pontificate of Alexander VI, who died in 1503, as well as those of his immediate successors, including Julius II and Leo X. That one could buy passage out of purgatory for oneself or others with indulgences—that is, by buying divine forgiveness through donations or "good works"—was particularly objectionable. In fact, Leo's sale of indulgences for the rebuilding of St. Peter's Basilica was the spark that lit Luther's ninety-five "fires."

Eventually, the general response of those who remained loyal to the Catholic Church was to meet the reforming northern protesters halfway—that is, to meet them with the Counter-Reformation, in which the "this world now" emphasis was disciplined and the balance of concern tilted back toward a preoccupation with life after death. As a consequence, much of the 1500s and 1600s saw European culture preoccupied with anxiety over personal destiny after death—to a degree perhaps unequalled in human history.

John Calvin believed with Luther that "good works" could not possibly save souls and asked how any act by a mere human could buy God's friendship. In effect, he asked: "Can one bribe God and thereby negotiate someone's quick entry into heaven?" His resolute answer: "No. If a human ends up in heaven, surely it can be only because God's graciousness has put them there; nothing in one's personal activity could have the efficacy to do so."

So Calvin not only affirmed what Luther said but then—more emphatically and single-mindedly than Luther—took a further step: God's efficacy is total, and because God causes *everything*, humans do not cause *anything*. God, he taught, can never be surprised or disappointed or defeated. According to Calvin's teaching, refined by his disciples soon after his death, the minutest details of the physical universe have all been preordained. It follows, then, that everything in existence has precisely the character it does by reason of God's eternal decrees.

No exception was made for human beings. If people were good, it was because God made them so—and the same held true for those who were evil. From the Old Testament and from Paul's reflections in his Epistle to the Romans and elsewhere—where Paul explains that God chooses some people for salvation and others for judgment—Calvin concluded that our race is a "fallen" one. While some of us are called out of that state and taken up into friendship by God, the more strict Calvinists believe most members of our species are "vessels of wrath," destined by God to burn forever in hell. To this day, such Calvinist Puritans believe only a minority are "vessels of grace" destined to be forever with God.

It's important, though, to realize that Paul's writings also insist on human responsibility. Throughout his letters, Paul urges people to repent, to believe, and to live according to God's will, suggesting that humans still have meaningful freedom and moral agency. Calvin's interpretation, however, narrows this space, presenting a vision where divine predestination overrides human choice.

If one wishes to understand the sense of superiority and American exceptionalism we bring to the conduct of foreign policy, this is a major clue. What is important to grasp is how—in processing the anxiety generated by Calvin's doctrine of inscrutable divine power—we Americans are endowed by a mandate to govern the world. Until recently, we've comforted ourselves by forming alliances with those whose prosperity we take as a sign of election.

In this theology—far more indebted to Calvin than to Jesus—the German sociologist Max Weber finds that God's ways become unsearchable and irreproachable, and there is no way for a person to know here and now whether they are a vessel of grace or a vessel of wrath, no way for a person to know at present what is going to happen to them for all eternity. Further, there is nothing that they or the Church or God Himself can (or will) do to change things. God's decrees are eternal and immutable.

Weber underscores how radically this Calvinist doctrine diverges from the New Testament. He writes:

> The Father in heaven of the New Testament, so human and understanding, who rejoices over the repentance of a sinner as a woman over the lost piece of silver she has found, is gone. His place has been taken by a transcendental being, beyond the reach of human understanding, who with His quite incomprehensible decrees has decided the fate of every individual and regulated the tiniest details of the cosmos from eternity.

In this doctrine of the early Calvinists, Weber finds a kind of ironic triumph of rationality over human sensibility and experience. If predestination is accepted, this doctrine cancels the teaching of Jesus both as a *summons* to freedom and as a *promise* of freedom. Predestination denies that we humans have any autonomy. We are not agents. That density of experience in which we feel we are acting as agents, taking charge of ourselves and altering the course of events, is regarded by Calvinists to be delusional. Over time, devout and committed Puritans come to regard all their experience as prefabricated. Apparently, we are mere spectators as one inevitability follows another. In such an understanding, our sense of paralysis when confronted by global warming becomes, it seems, an exercise in theological realism. One feels as though one is a mere puppet, and that one's sense of Nature as a realm one can enter to pursue a dialogue with Nature's Author is an illusion.

Calvin's quest for consistency installed a capitalism by which we are now enthralled. Central to Weber's book *The Protestant*

Ethic and the Spirit of Capitalism is a boldness of statement that probably didn't come easy for Weber. He is saying that howsoever sincere the Calvinists may have been, their doctrine of predestination places their disciples in a quandary alien to Christianity.

Chapter 7

Puritanism in the New World

IN 1630, A GROUP OF PURITANS, including John Winthrop, a lawyer, left England and journeyed westward across the Atlantic to help establish "New England" in territory traditionally held by the Massachusetts tribe of Indigenous Americans.

These Puritans did not hold the Anglican beliefs that were maintained after Henry VIII declared English religious practice severed from allegiance to Rome. Massachusetts Calvinists regarded Henry, his daughter Elizabeth, and the Stuarts who succeeded them as unreliable dissenters from Roman corruption. These Puritan Calvinists felt there was too much remaining that was sympathetic to "Romanizing" elements in the structure of Anglicanism. These elements needed to be rooted out and replaced by the "pure" Christianity that Puritans promoted.

While they distanced themselves from the heavy hand of the English monarchy, these Puritans brought with them the conception of a heavy-handed God from which there could be no escape. In their newfound physical independence then, how were they to go about performing the routine tasks of everyday survival while burdened with the thought that they might soon be irrevocably damned and begin to burn eternally in hellfire in the life to come?

For Max Weber, their anxious concern provides a clue to the sturdy presence of a capitalism, whether religious or secularized, that two generations later economist Milton Friedman would advertise as the adequate foundation for all ethical behavior and all organization of American life and purpose. It is a surrounding and engulfing capitalism on behalf of which we now routinely send our best and brightest overseas to fight and perhaps to die, saying that such is the sacrifice we make for the preservation of American freedom.

While Weber paid little attention to the lives of individual American Puritans, the suggestion he provides for understanding them is that the immense burden they carried with them across the Atlantic was unbearable. In the days of New England's founding, Puritan pastors were beset by the urgent need of their flock for relief from anxiety. As was the case in Europe, the relief the pastors eventually provided was a kind of reworking of Levitical spirituality as featured especially in Deuteronomy. There we are told that the good man prospers and the evil man fails.

The Puritan pastors, feeling an obligation to give reassuring counsel to their layfolk, reworked their spirituality by placing emphasis on this principle. While it is true, they argued, that good works (such as purchases of indulgences) cannot win God's favor, is it not likely that success in one's worldly affairs here and now was a sign that one is among the minority God has predestined for a life of eternal happiness? In fact, is not the reassurance to be gained from such reflections a reassurance that will expand and deepen during a life of continuing prosperity?

Surely the greater one's success in the everyday world, the more secure one can feel regarding one's election to grace. And as it is clear there is no such thing as too much reassurance, so too it seems there can be no such thing as too much success. Thereby was the heavy hand of God transformed into a steadying and relentless divine push toward relentless activity and self-aggrandizement. Though the phrase was not yet in common usage, the notion of the "prosperity gospel" had taken root.

From this notion Weber anticipates the transformation—not complete in his day—by which capitalism becomes an ethics unto itself, an ethics endorsing an autonomous economics capable of becoming (as Weber muses toward the end of *The Protestant Ethic and the Spirit of Capitalism*) an "iron cage." Iron because its adherents regarded it as something dictated by God, and a cage because it excludes all other concerns. Weber regarded it not as a system we take in hand so much as one that has taken *us* in hand.

He saw Benjamin Franklin as a prophetic figure, an instantiation of what was to come. In Franklin's elaborations on "a penny saved is a penny earned," Weber saw the future of America. When, over time, the religious preoccupation dimmed (as it seems to have dimmed in the personal life of Franklin), the addictive pursuit of the "signs of success"—like a gambler's addiction to risk—would endure and grow. Milton Friedman, one can argue, consolidates this ascendancy by uncritically relegating everything except the acquisition of money to the status of an "externality."

The tendency of this simplification is to leave us no other way of keeping score, no other way of finding meaning. This is America today. As more than one commentator has remarked, American businesspeople seem able to imagine an end to the world, but most of them seem unable to imagine an end to capitalism. Businesspeople draw consumers into an endless quest for "security"—a quest that is now despoiling the planet.

Those who accept Weber's arguments recognize we are descended from a traumatized people. The Calvinist framework for our trauma may have faded from memory for many of us, but in the last half century—as American Catholicism has merged with Puritanism in upholding capitalism (see the strong Catholic presence on the Supreme Court), there has been a remarkable intensification of religiosity. This is now central to the ideology of the Republican Party—and also to much of the Democratic Party. To many Americans, Donald J. Trump seems the living incarnation of "success"—a striking realization of the prosperity gospel, a new Moses restoring us to the Promised Land.

It is not extravagant to say that Weber sees our brand of unbounded capitalism as a product of religious neurosis. Most of our pursuits have come to be organized around and guided by the pursuit of money and its accompaniments. While this tendency has been operative throughout human history, perhaps only now has it been released from nearly all constraints imposed by empathy toward the neighbor or respect for the symbiotic structure of the natural world in which we've emerged as dominant. Like Cato's version of Carthaginians, we sacrifice our young to Moloch. Think too of pathological "stand-your-ground" authorizations to kill one's neighbors in some states, and of the freedom with which some of us attack entire species of wild animals.

Theoretically, we can be the preserver or the destroyer. In actuality, we seem to have appointed ourselves the superconsumer, grabbing with uncensored recklessness any riches the earth has to offer at the present moment.

Weber recognizes that capitalism is a resolution of sorts, for it reassures us and fills the void of uncertainty in which we might otherwise stand. Yet he sees it as a resolution deeply rooted in drivenness and the surrender of personal dignity and freedom. He writes of the Puritan: "The idea of a man's duty to his possessions, to which he subordinates himself as an obedient steward, or even as an acquisitive machine, bears with chilling weight on his life."

The word *steward* should not confuse. We are not reading here about service to nature. Weber writes of "the entirely negative attitude of Puritanism to all the sensuous and emotional elements in culture and in religion." While Weber was inclined to think a sociologist should refrain from positive and negative evaluations in descriptions, the explicit negative judgments found here indicate that, as an actual writer and living human being, it was impossible for him to maintain such detachment. Just as a good physician comes forth not merely with a theory of the human body but also with empathy and a diagnosis for what's wrong with a particular patient, so too does Weber with regard to the emerging world.

The plausibility of what Weber observes but doesn't endorse can be monitored in our continuing sense of American exceptionalism. From its perspective, we are an "elect" people, but no freedom of ours plays a part in our election. We haven't volunteered to be the Elect. It is imposed upon us. We are Chosen.

John Winthrop envisioned us—then Puritans, now Americans generally—as "a city upon a hill," and we have uncritically accepted his vision. Here we are, manifesting God's glory to the world at large. We provide Liberty and Justice for All. We have a Manifest Destiny. We are humanity's Last Best Hope. We are divinely commissioned to rule. That's why whoever we elect president becomes thereby The Leader of the Free World. With ardent [Woodrow] Wilsonian idealism, we fancy ourselves as protecting the right of people everywhere to self-determination. The only restriction we place on this right is that they exercise their self-determination in a manner that we can recognize as conducive to our success and God's triumph.

THE FOUNDING PURITANS of Massachusetts found themselves engaged in an "errand into the wilderness"—a phrase twentieth-century Harvard historian Perry Miller found meaningful in his discussions of them. Some Puritans would eventually return to England to assist in the 1649 beheading of Charles I, who was perceived as too tolerant of dogmatic and liturgical contaminations and too ready to engage in diplomacy with Roman authorities. The majority of Puritans, however, remained in New England as a "body of the Elect" who saw themselves as conquering for God an "unpopulated wasteland."

The last two words translate a phrase associated with Puritan John Locke's musings on property: *vacuum domicilium*. Whether Locke actually used the phrase is not a matter of great importance; he articulated the context. The land was "unpopulated" or "vacant" because Indigenous Americans didn't count.

Puritans conveniently "discovered" that the Indigenous tribes made no "use" of the land and had therefore no "civilization." From this lack, one could know the natives weren't among the Elect. They were, like most of humankind, destined for hell. In the meantime, they were something close to sheer obstacle. As Teddy Roosevelt later assured us in *The Winning of the West*, a death among them was a victory for righteousness. That the wasteland was bountiful, and that our success in claiming it was total, proved that our victory gave glory to God just as it gave us assurance that we enjoyed God's blessing.

If, as suggested, we look at our current foreign policy, we can recognize how much of the Puritan mindset continues to set a pattern for current behavior. As descendants of Puritans (by naturalization if not by birth), we came by virtue of the Monroe Doctrine to recognize that the whole of the Western Hemisphere was our legitimate sphere of influence.

While this perspective was dawning, it was easy to deduce that it was our destiny to incorporate northern Mexico into the United States. Earlier than that, in the time of the Spanish-American War, it seemed evident to us that Cuba was a "fruit waiting to fall" into our welcoming hands. Defeat of the Spanish was validation. In vanquishing them we made Washington, DC—not Madrid—the deep source of Cuban law. A couple of generations later, under Castro, when many Cubans rejected our hegemony, we put the future of the entire planet at risk rather than accept such a violation of our authority.

Globalizing the Monroe Doctrine, we currently find it is up to us to decide which nations shall be permitted to have nuclear weapons. At the same time, we find it absurd that other nations should have a say in whether *we* should have them. Not long ago, we regarded it as a part of our God-given mission to straighten out Vietnam. The fact that this turned out to be a merciless business didn't deter us in the following generation from taking on the Middle East as a God-given American responsibility: *Their* self-determination was not being exercised to *our* satisfaction.

Chapter 8

Puritanism Meets the Enlightenment

LET'S REFLECT ON OUR USE of the phrase "father of our country." The standard way young Americans get introduced to the colonial origins of our country conflates Puritans with Pilgrims and memorializes the first Thanksgiving as a rose-colored lens through which to view how our colonial ancestors interacted with Indigenous Americans. The colonists present at that Plymouth Thanksgiving were actually a subgroup of Puritans called Pilgrims. They were mostly Calvinists who wanted to be left alone to practice their beliefs and who did not regard themselves as warranted to execute Squanto—a member of the Patuxet tribe who helped the Pilgrims survive and acted as a liaison for them—or other Indigenous Americans.

The next wave of Puritans, by contrast, did not want to leave anyone alone; they wanted to take the world in hand. (In the southern colonies, Thomas Jefferson, James Madison, and James Monroe would come to show their affinity with Puritan ideology in general, and with Locke in particular, by their conviction that they owned great masses of fellow humans captured in Africa and by their belief that land occupied by Indigenous Americans was "unoccupied.")

The Puritans of Massachusetts Bay Colony arrived ten years later than the 1620 *Mayflower* and its Pilgrims, and they were not present at the Thanksgiving after the 1621 harvest. Yet John Winthrop, leading Puritan and multiterm governor of the Bay Colony with "city upon a hill" ideals, seems to have a strong claim to be the founding father of present-day America. (His absence from the first Thanksgiving seems appropriate, as he would not have fit into that idyllic picture.)

Complicating the teaching of American history is that we have fixed inaccurately on a supplementary "father of our country"—George Washington—even though this father is not only chronologically second to Winthrop but also contributes less in a genetic way. By that, I mean to say that when there is conflict between the legacies of the two men, Winthrop's legacy prevails.

We think of Washington as molding the country, but a more accurate view might be that the country molded him. Anglican that Washington was, much of the mold into which he was cast was shaped early in the hundred years preceding his birth in 1732 and has shaped who we Americans are today. Our immigrant ancestors regularly came here to surrender their previous identity and inherit the Puritan legacy.

The minority of historians who don't see Washington as our first father are nearly excluded from the conversation about American origins. The excluded minority have a right to complain of the kidnapping of American history by the pseudohistorian Parson Weems, Washington's first biographer, and by Weems's recurring and predictable disciples. Weems did not dedicate himself to biography but to hagiography. He flatters our spirit.

He could not have achieved this had Washington not had qualities apt for legendary elevation. While Washington may not have been the perfect man we conveniently commemorate, it is obvious that he did have a steadiness and objectivity evident to all but those blinded by envy. Specifically, he carried as a kind of birthright the aura of the Enlightenment that became so important to us from at least the middle of the 1700s.

We like to think of our country as the first nation rooting itself in the clarity of the Enlightenment. That movement arose in Europe in a resolute effort to transcend inherited privilege, religious bigotry, and violence, and to embrace the open-ended opportunities for progress that science was making available. In France, its spirit was expressed in the motto *Liberté, Égalité, Fraternité*. In Germany, it found expression in the ethics of Immanuel Kant: "Never treat humanity—whether one's own or another's—as a mere means, but respect it always as an end." People of British ethnicity on both sides of the Atlantic tended to regard this spirit as the culmination of a long quest toward recognizing the "rights of man"—a process they viewed as dating back at least to the days of the Magna Carta (with some thinking the inspiration preceded the Norman Conquest and was indebted to the legendary Arthur and his Round Table).

The ideological tone of the Enlightenment was radically different from that of Puritanism. While there were theists (believing in a god active in the world), atheists (no god), agnostics (undecided), and deists (a creator god *not* active in the world) among the figures of the Enlightenment, the leading intellectuals were likely deists. Their God was one who had established the context of an ordered universe ruled by gravity. This God was believed to have fertilized matter with the seeds of a vast diversity of species that interacted on each other synchronically, and then had pretty much left this planet to the care of us humans, endowed with freedom and the creativity to do as we saw fit.

The idea of an Elect among humankind was alien to this Enlightenment with its tendency to proclaim the equality of all. A notion then that most of humankind were damned wasn't just repugnant to Enlightenment thinking but was dismissed by the intellectual elite as ridiculous. And if one were to carry through the critique of privilege that was a central theme of the Enlightenment program, one had also to reject predestination and divine preferences. Paradoxically, the critique of privilege meant the Enlightenment would give rise to strife. Intellectuals of the Enlightenment didn't see why the mere fact that so much had been owned by a minority meant that things should continue to be that way. As a result, they found themselves in opposition to the Puritans.

Particularly in France, humanity came to be regarded as composed of individuals of equal value, each of whom had all the birthrights of any other human being. Manifestly, this was a revolutionary notion—one that invited people to take issue with many of the traditions and institutions into which they'd been born. Notions of indelible claims on property, for instance, could not easily be reconciled with advocacy of liberty, equality, and fraternity. (Jefferson, in this regard, was a wistful dreamer. He liked the idea of an Enlightenment so long as it remained thousands of miles away from his plantation, which was replete with enslaved persons.)

In England and Scotland, as in France, the Enlightenment's free spirit of scientific inquiry inspired in the political order an effort to reduce the influence of religious denominations—often presenting reform as an effort to restore guarantees of individual rights thought to have been observed in simpler times. (In mainland Europe, the Treaty of Vienna of 1815 seems to have had a tranquilizing effect.)

What had happened earlier in France was that the ideas promoted by an intellectual elite led to an explosion of murderous enthusiasm aimed toward replacing the most familiar arrangements of French society. We call this the French Revolution—and when we want to denounce it, we speak of it as the Reign of Terror.

Regarding America a century earlier, one can view the arrival of Puritans as bringing to America a pre-Enlightenment critique of Europe, but this critique should not be conflated with the Enlightenment itself. The specific character of the American Puritan critique was, as said, a result of a quarrel in England regarding Anglicanism, Romanism, and the English throne. The Stuart kings were regarded by Puritans as Catholicizing aristocrats—imposters from whose hands the true faith of Christianity needed to be wrenched.

In the rare moments when John Winthrop and company have been presented in our history courses, it has been a somewhat routine mistake for them to be portrayed as champions of religious freedom. To regard these forebears as people who sought nothing more than to be left in peace to pray as they saw fit is to neglect most of our actual data on them. They did not regard themselves as champions of religious freedom but rather as champions of Calvin, whom they saw as a besieged champion of Christian teaching. Their move to New England was a maneuver in the long game they were playing for the soul of British spirituality. Historian Perry Miller, taking his cue from Winthrop, suggests that they intended to reform European spirituality in general by the example they set in Massachusetts Bay Colony.

As mentioned, Puritans who journeyed back to England during the early years of the colony labored successfully for the execution of Charles I. Intolerant of the "impurities" in Anglicanism, they were intent on getting rid of them. Their "errand into the wilderness" was a kind of hunkering down and regrouping in the New World in preparation for taking firm hold of the old. This "internationalist" perspective has been with us ever since.

To sharpen the point, let me repeat that there is hardly a better formula for intolerance of others than the Calvinist doctrine of predestination. And, as said in an earlier chapter, the Calvinists had—and still have—an opportunity to claim biblical authorization. An instance not yet mentioned: In the Old Testament, predestination functions as the premise for the book of Joshua. In that superhero fantasy book (one for which the commentary in my copy of the Hebrew Bible says there is no historical evidence), the annihilation of Canaanites is not portrayed as punishment for Canaanite sins—it is presented as fulfillment of a divine command. The Israelites were *commanded* to invade Canaan. To use the vocabulary later in use among Puritans, the Canaanites were Old Testament "vessels of wrath." They had been created to be destroyed. It is equally obvious in the text that the Israelites—taken in hand by God as divinely crafted weapons—were "vessels of election." Both groups were destined by God for the roles they performed.

Paul's argument for God's absolute sovereignty brings predestination into the New Testament. Although Paul seeks to preserve human freedom through his calls that the Christian choose to imitate Christ, there seems some dent in his philosophy of freedom. Paul's argument for God's control of history is distilled to a pure dose by Calvin, and we seem to become cogs in a giant machine run by God.

While the deep theory of predestination comes in fact with some frequency to the fore in all three Abrahamic religions, I'm inclined to agree with some Reformed rabbis that wherever one finds it, it is there as a devout figure of speech—an act of conscious hyperbole. One can say that to the extent it is legitimate, it is there to say that God's providence is all-embracing—that God, like a good parent, has purposes and makes provisions for the human species from the beginning to the end of human history.

In the twentieth century, a temperate theology of benign parental providence was the view of such theologians as Dietrich Bonhoeffer and Martin Luther King, Jr., and of the British philosopher Alfred North Whitehead, whose father was an Anglican minister. While these men showed a lively sense of the actuality of evil, each felt sure evil was not God's doing; and while they believed God cared for the universe, they denied that God micromanaged our use of it in our acts of freedom. They thought that if God were stage-managing each moment in human affairs, our lives would not be our own—and, presumably, Hitler could not have happened because God would never have caused him.

The predestination of Calvinism is a drastic theory of providence, and Calvinists don't see it as metaphorical. They move quickly from a premise common to Judaism, Christianity, and Islam that God is *present* to human affairs, to an affirmation that God is the *cause* of whatever happens. They then pass to what seems to them a logical corollary. As mentioned earlier, if God is the cause of all things, humans have no causality at all.

If, in company with some twentieth-century existentialists, one affirms that "to exist" equals "to be capable of agency—to be capable of *causing* things," then you might say Calvinism disallows the very notion of creation. For we are not then distinct from God; rather we seem to be mere images in a game of divine solitaire. Calvin (without intending it) seems to reduce reality to the reality of God.

This is not compatible with the teachings of Jesus as presented in the Gospels. What Aquinas, Bonhoeffer, King, and Whitehead regard as "creation," with its vast tapestry of cause-and-effect relationships, gets devalued into a charade—which may help explain why so many American descendants of Puritans seem indisposed to take seriously the version of nature that science investigates and reports on. To say humans are causing climate change is illogical if in fact humans never cause anything.

Sin becomes particularly problematic within the spiritual economy of Calvinism. One can say, as Calvin did, that the doings of God are inscrutable. The term *sin*, though, is a human invention; we know what we mean by the term. Its meaning is *not* inscrutable, and it implies freedom. (For this reason, in our jurisprudence, we have available the verdict "not guilty by reason of insanity." Presumably the insane lack freedom, and to the extent they lack freedom, they are regarded as incapable of sin.) Once freedom has been subtracted as a condition for the notion of sin, one must wonder what sense of the term remains.

If God causes all my acts, then whether they are acts of virtue or acts of sin, they are all equally acts of God and none of them is really mine. In any and all of my acts, I do neither more nor less than what God wills. But if sin is fully conformed to the will of God, one must wonder how it differs from other acts that are fully conformed to the will of God. What gives it its character as "sin"?

In such an arbitrary view, there is no empirical criterion by which we mortals can distinguish good and evil, and the consequence for human behavior is that it is likely to become unruly and terminally destructive. As we will see, this was the case for the inhabitants of Massachusetts Bay Colony, and their behavior toward Indigenous Americans became a prototype for later generations in our behavior toward Mexicans, Cubans, Filipinos, Russians (at the end of WWII), Vietnamese, and Iraqis. And now Venezuelans.

In prevailing modern cultural attitudes, the term *evil* lacks an objective correlate—that is, evil is seen as subjective and no shared moral standard exists. In discarding the term, however, we jeopardize civilization and move rapidly toward rendering the planet uninhabitable. Considering the ease with which we say "anything goes," it's difficult to maintain optimism that our species will still be functioning in a civilized way at the end of this century. Chaos seems the more likely outcome.

Chapter 9

The Puritan Precedent Set at Mystic Fort

ROUGHLY SEVEN YEARS after Winthrop's June 1630 arrival at Massachusetts Bay Colony, uneasy relations between the Puritans and the Pequot Indians erupted into what has sometimes been called (by those aware of it) the Massacre at Mystic Fort or the Pequot Massacre. While we have documentary accounts of events leading to the massacre as well as of the massacre itself, the subjective character of these documents, written by Puritans, leaves room for interpretation. Some framed the massacre as the result of a series of bloody encounters, and others framed the Pequot as pure victims who were obliterated in a single onslaught.

At the start, the Puritans do not seem to have had an explicit genocidal agenda toward the Pequot. Rather, what the Puritans seemed to want was hegemony. The Bay Colonists wanted as much control over the fur trade as they could get and evidently believed they could achieve this by gaining the upper hand in their relationship with the powerful Pequot. By 1636, the Bay Colonists also probably wanted to maintain the upper hand with the fledgling Connecticut Colony of Puritans to the southwest. Winthrop and company may have felt that maintaining control of the Pequot would help in keeping a measure of authority over the Connecticut Colonists. Further, as the Bay Colony grew in population, there was likely a desire to extend established boundaries—a concern that would have made neighboring Indigenous tribes rightfully ill at ease.

By 1636 then, there's reason to think the Bay Colony was looking for a *casus belli*—an excuse for war—in regard to the Pequot. Earlier, a bad actor named John Stone had been expelled from the Bay Colony. Another, John Oldham, had been expelled by the Pilgrims but seems to have remained useful to the Bay Puritans. Eventually both men were killed by Indigenous people, though neither seems to have been killed by members of the Pequot tribe. Nonetheless, the Pequot were thought to be providing haven for individuals from the Niantic and the Narragansett tribes who were suspected of involvement in the deaths of these White men.

In 1636, the Puritans told the Pequot that the surrender of the suspected killers was a condition for continued peace. This looks opportunistic, considering Stone had been an outcast. For the Puritans to abruptly remember him as a fellow citizen suggests that most any provocation would have sufficed. The Pequot, wishing to maintain peaceful relations, offered some wampum (a traditional shell bead used as a medium of exchange) in compensation for the deaths of Stone and Oldham, but said their own alliances with tributary tribes did not permit them to hand over the suspects.

One should recognize that the Pequot reputation with other tribes was at stake; they had not gone out of their way to insult the Bay Colony. It seems it was the Bay Colony Puritans who issued the ultimatum.

According to one account, at dawn on May 26, 1637, Puritans of the new Connecticut Colony joined up with a lesser force from the Massachusetts Bay Colony and attacked the Pequot at Mystic Fort, the Pequot's fortified village. Puritans entered the palisade, set fire to the dwellings, burned alive every man, woman, and child they could find, and posted themselves at the two exits of the palisade to kill anyone who tried to escape. (This was an enactment meant to take the Battle of Jericho out of its mythical setting and make it real.) Somewhere between four hundred and seven hundred Pequot individuals are thought to have perished that May morning. One account says that at most about fourteen Pequot escaped death, and in the aftermath, they were hunted so they, too, could be killed.

John Mason, who commanded the Connecticut contingent at Mystic Fort, described the events as follows:

> Thus were they now at their wits end, who not many hours before exalted themselves in their great pride, threatening and resolving the utter ruin and destruction of all the English, exulting and rejoicing with songs and dances. But God was above them, who laughed his enemies and the enemies of his people to scorn, making them as a fiery oven: Thus were the stout-hearted spoiled, having slept their last sleep, and none of their men could find their hands. Thus did the Lord judge among the heathen, filling the place with dead bodies.

Notice the action was not recounted as something John Mason and his command had done; it seems God did the whole thing. This is the convenience of predestination—no one among the Elect can be blamed. A transparent anomaly in Mason's account is that it has the Pequot exalting themselves in their great pride yet making no defense of their home. The success of the attack required, in fact, that the Pequot be asleep when it began.

The Puritans established here an American precedent that has not since gone out of style. As Winthrop and company did in their day, so do we in ours advise foreign peoples against what we would regard as an unneighborly course of action. If they accept our advice, we may spare them from negative consequences. If they do otherwise, we confront them with reprisals—reprisals often culminating in annihilation.

Body counts from war are difficult to ascertain. Does anyone have a solid figure, for instance, for the number of Indigenous Americans killed in "winning the West"? In our tally, do we count the deaths of native peoples from hunger and malnutrition due to the purposeful extermination of buffalo herds? In the case of Vietnam, deaths of Vietnamese from shooting and bombing by Americans are listed by some as just under a half million and by others as well over a half million. But then we have been told by Vietnamese officials that five hundred thousand Vietnamese children have been born with birth defects resulting from the war. Are we willing to count the children who died from these defects as children killed by us? (*Why should we?* we reason. *After all, many of these children were the children of communists.*) Further, where we bombed dikes, do we count those who drowned soon after? Do we include those who eventually died as a consequence of floods, disorder, and hunger? (Three-time veteran Jim Homan, whose ashes I helped spread off Guam at Tumon Bay, thought we should.)

If to our Vietnamese victims we're willing to add tallies from the Spanish, Cuban, Filipino, Mexican, Iraqi, and Afghan aggressions we've initiated, it is likely our victims run into the many millions. Responsibilities for current wars in Ukraine and Palestine perhaps remain less clear.

Although Russia, China, Pakistan, and India all have nuclear weapons, we continue to issue threatening demands—even in the full knowledge that if we or they resort to nuclear weapons, it's quite likely such action will trigger nuclear activity from others and lead, as a consequence, to the onset of radioactivity throughout the planet and the corresponding loss of habitat for our species.

The contest between celebrating creation as the Enlightenment invited and exploiting creation toward a point of ultimate ruination as scorched-earth capitalism does has never been more urgently present than it is today.

We argued in our Declaration of Independence for our right to discard the rule of England. We did so with an argument from natural law regarding the universal rights of human beings. Yet through two and a half centuries, we've never been wholehearted in implementing our Enlightenment commitments. Judged by his own words, the man who wrote high ideals into our Declaration of Independence was a self-contradicting slaveholder. His political career was largely in the service of a desire to keep our federal government too weak to deprive his class of the comfortable lifestyle they had as beneficiaries of slave labor. He was, and remains, one of the most revered characters of the American story.

From the Revolutionary War on, it has been our custom as Americans to use the rhetoric of the Enlightenment to justify whatever it is we want to do. Reaching back to Winthrop, we've said we want to be "a city upon a hill"—a source of liberty and justice for all. What we've routinely done, though, has been to behave as our forefathers did toward the Pequot. We see this legacy in the lives of Jefferson and Madison, Monroe and Jackson. Or rather, we see it if we *consent* to see it. These men knew quite clearly that they were exploiting the lives of the Black people they held captive. They knew the talents of these captives, and they knew they could not have the lifestyles they were enjoying were it not for the extensive skills and hard labor of people they pretended were inferior.

These American politicians weren't insane, but to understand them perhaps it's necessary to let Jean-Paul Sartre help us with his reflections on bad faith. Sartre, if I have understood him, would have it that our human freedom is of such power that we can simultaneously know and not know that some inconvenient truth is present. That is, we can deceive ourselves. At some level of consciousness, our American patriots must have known from the tops of their wigs to the soles of their buckled shoes that they did not take seriously what they said about human equality.

We must take note of this doublethink, for we continue to say one thing but behave in another way. Many of us have "bad faith" for breakfast every morning. We view our Bibles through Calvin's eyes and find ourselves revealed as inheriting the birthrights of God's chosen. American Catholics every bit as much as American Protestants have seen themselves as the Puritan Elect. As Weber was inclined to expect, a majority of us reconfigure the world about us as a world of Canaanites. We are tempted to see everything—and sometimes everyone—as ripe for the taking; and then we bless our holdings with pious statements about the equality of humans.

The politics generated by this tension between the Enlightenment and Puritanism is interesting but not well understood. We've come to regard Jefferson and Jackson as champions and liberators of the common man—patron saints in our American version of "liberty, equality, and fraternity." We pride ourselves on our insight into human equality and feel confident this insight makes us far superior to others.

And yet Thomas Jefferson was not above complaining that having slaves was hard work but someone had to do it. The principal harvest at Monticello was, in fact, the children of captured Black people. Jefferson used the increase in his slave population as collateral for ongoing borrowing to support his otherwise unaffordable lifestyle. When he died, the families he had exploited throughout his life were split up and sold off in different directions to make some payment on his debts. It seems only the Hemings family was fully spared. While he liked to fancy himself a simple farmer, it has been a rare thing in human history for one man to have enjoyed such intimate and absolute personal control over as many lives as Jefferson did. Among those who did rival Jefferson were his White plantation-owner neighbors, and he lived in service to the class system that supported their lifestyle. His *Notes on the State of Virginia* makes it clear he considered himself among the Elect. He took it as a sign of providence that slaves were available at just the time he most had need of them. Winthrop never had a more fervent disciple.

As for Andrew Jackson, while he may not look the part of a Puritan, when he wrote to President Monroe about treaties he'd crafted with Indigenous Americans to ease them off their land, he told Monroe he felt such treaties imposed no obligations on him or the American government. Later, when he himself was president, he sponsored the removal of the Cherokee and four other Indigenous Nations from their farmlands in the Southeast—a removal remembered as the Trail of Tears and carried forward by his successor Martin Van Buren. When John Marshall as Chief Justice of the Supreme Court affirmed a treaty right of the Cherokee to remain on their land, legend has it that Jackson said sarcastically, "John Marshall has made his decision; now let him enforce it." Whether or not he actually said those words, Jackson's intention was to free up lands for plantations whereon Blacks would do the heavy lifting. He certified himself as a dutiful Puritan by believing American Whites had been elected to a unique status. It was for this that he was—and remains—popular.

Any genuine reverence for Enlightenment thinking in our early days as a republic was limited to a number of Federalists. Foremost among them were Alexander Hamilton and—though he abandoned the Federalist label—John Quincy Adams. One should admit that these leaders, too, had an aristocratic sense of themselves, yet notions of freedom and equality weren't for them mere rhetorical puffery; they were the astronomical presences that lit their lifestyle. (Their party did not endure. The quixotic candor of the party's creative genius, Alexander Hamilton, had much to do with this.)

One could claim that federalist thinking made a brief return in Lincoln's administration. A dozen years after Lincoln's death, however, one can see that Republicans abandoned campaigning for Black enfranchisement and left the Blacks of the South (and elsewhere) to whatever mercy the majority White populations were willing to concede. The Democratic Party of Jefferson and Jackson became the unapologetic curator of Jim Crowism—a program of denying Black people rights and opportunities regarded by Whites as God-given and reserved for them alone.

Woodrow Wilson, a Virginian elected by reason of a falling-out among Republicans, may have sounded like he revived the smoldering embers of the Enlightenment, but what undid him was his allegiance to Winthropian arrogance and the Puritan legacy. The liberty he endorsed was the liberty of other people to do what he wanted them to do. That that did not always happen did not redeem his presidency.

We had a "return to normalcy" under Warren G. Harding, which meant that an unshackled Puritan acquisitiveness reigned until the economy imploded in the Great Depression.

This was the state of things when Democrat Franklin Delano Roosevelt arrived at the White House in 1933. With his presence, the rhetoric of the Enlightenment—supported by federalist political thinking—made a comeback. But Jefferson and Jackson still had places of honor at the party table, and FDR had no desire to lose the South to Republicans; he needed Southern votes if he was to stay in office. One can plausibly endorse the argument of Stanford historian Barton Bernstein that FDR did not so much reform capitalism as preserve it for new ventures after his death.

As Wilson had his Harding, so Roosevelt had his Harry Truman—from Missouri, tutored by Churchill, and pedaling backward quickly from the more humane aspects of Roosevelt's foreign policy. We would force-feed Western Europe on capitalism and do all we could to intimidate the Soviet Union. This position not only registered a lack of gratitude to Russians for sacrifices they had made to defeat Hitler's regime, but it was not good domestic politics. Think of the confusion when the

Democratic Party, inheritor of Enlightenment rhetoric during the New Deal (the Four Freedoms and so on), routinely conducted fundraisers in honor of Jefferson and Jackson, two of the most anti-liberal presidents in our history. It suggests that in neither party could we ever get far from Calvin. The rhetoric of both parties has, until recently, been the rhetoric of the Enlightenment, yet at step after step, our policies have routinely been dictated by our Puritan mindset.

Two surviving children of an Iraqi family killed by American soldiers (Photo: Chris Hondros/Getty Images). As the sun set on January 18, 2005, an Iraqi family was hurrying home from a visit to a hospital. At 5:50 p.m., they were close to home and had ten minutes in which to meet the curfew, announced in English, that we Americans had proclaimed in their homeland. Since their car didn't slow down when it approached a patrol of American soldiers, the soldiers evidently decided "better safe than sorry." The mother and father were killed almost immediately. "In a perfect world, this would not have happened, but we do not live in a perfect world," a member of our military thoughtfully explained.

Chater 10

The Emergence of a Secular Puritan Economics

IN EARLIER CHAPTERS, we noted a bifurcation in the foundations of American culture that has led to a dualistic character: We have rival traditions. We have competing and incompatible ideological premises. We simultaneously have John Winthrop and George Washington as our founding father. We have Calvinist distrust of this world as the devil's playground, and we have the Enlightenment's celebration of this same world as a place of shared light and opportunity for all in the pursuit of excellence. We inattentively regard these incompatible perspectives as providing one legacy. The Enlightenment perspective can, as noted, regress into paranoia and distrust—as France's Enlightenment morphed for a time into the Reign of Terror—but to see the two perspectives as one is a failure in critical thinking.

This lack of critical thinking (and our denial of it) has had immense implications for our politics. During the Second World War, when I was not yet a teenager, I frequently heard that we were traditionally a peaceful people, disinclined to the squabbles that kept Europe in constant turmoil. Within days of repeating that trope, we incinerated the centers of two Japanese cities, saying, "See what happens when a peaceful country like ours is provoked to war!" Truman said the American people would have thought him crazy had he *not* authorized the use of nuclear weapons against Japan. He regarded their use as common sense rather than aberration.

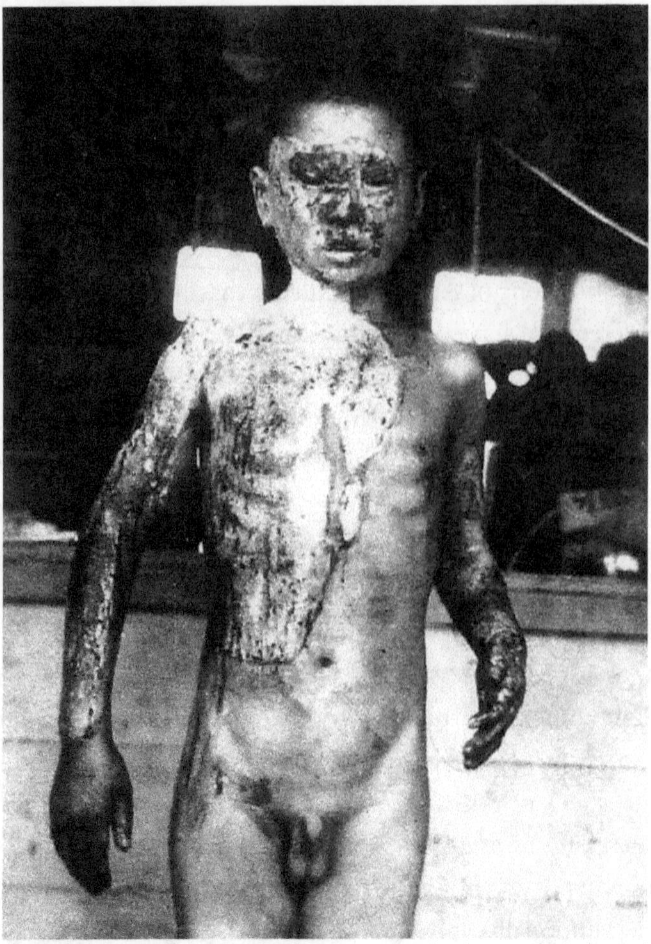

Radiation burns, Nagasaki (Photo: National Library of Medicine/ Science Photo Library). What were we doing here? Were we teaching this young man he should not have bombed Pearl Harbor?

We who had taken on the British, we who had abrogated almost every treaty we'd ever signed with Indigenous Americans, we who had invaded northern Mexico because we wanted it, we who had suppressed the liberties of Black Americans after the Civil War with Jim Crowism, including the normalization of extrajudicial killing (a.k.a. lynching)—we were *not* a peaceful people. For us, violent action functioned up there with apple pie and motherhood as a sacred icon.

We have noted that until quite recently we used the phrase "leader of the free world" to designate our president, and we still sometimes use "champion of the free market" to describe how we promote our economy worldwide. Notice, though, how dysfunctional our government has become. Our claim to be a democracy seems less and less supported by facts, and we take incoherence for granted. While most Americans say they prefer peace to war, our chronicle of events throughout the twentieth century and the first quarter of the twenty-first is a chronicle of one war after another. Another incongruity: Polls indicate that most Americans favor stricter government control over the sale of military-style guns to civilians, yet there is no progress in protecting children and other innocent parties from attacks with such weaponry. In yet another contradiction, most Americans favor government promotion of healthcare for our whole populace—the very nature of contagious disease makes this a matter of enlightened self-interest—yet resistance to maintaining movement in this direction is currently strong in what is nominally a "government by the people."

With regard to our wars, governmental mendacity is routine and consistent, dating back at least as far as James Polk's claim about the start of the war with Mexico and continuing forward with Teddy Roosevelt's claim that we "knew" the Spanish had blown up the USS *Maine*. In 1962, JFK was too ready to find a pretext in the Monroe Doctrine for an ultimatum to the Soviet Union that could have led to the annihilation of humankind. (One ought, though, to belabor this some. It's now clear to anyone still interested that author James Douglass was right in claiming Lee Harvey Oswald had been groomed by CIA handlers for years. Kennedy was assassinated by a coup involving members or former members of his own administration. During the Cuban Missile Crisis of 1962, he had arrived at an insight that the earth is not reducible to "America's sphere of influence." Key members or former members of his administration had not agreed, and he was deeply resented.)

Subsequently, LBJ lied about the Gulf of Tonkin incident in order to recruit support for our war against Vietnam. Later, G. W. Bush and company claimed evidence for nonexistent weapons of mass destruction in Iraq. Later on, Barack Obama defaulted on his promise to end the anomaly of keeping prisoners at Guantanamo without any prospect of due process. Obama proclaimed that violence never solves anything, yet he continued to authorize drone assassinations and sent Navy SEALs into Pakistan to conduct a nonjudicial execution of Osama bin Laden. In 2008, Obama campaigned as an agent of change, yet soon after he was elected, he decided to reengage in our counterproductive war in Afghanistan, with no clarity regarding what would constitute victory. It looks like he was merely leading us in a latter-day Wilsonian effort to teach Afghans how to be good Afghans.

The Enlightenment is still with us, but as I've remarked, it seems to exist throughout our history primarily as a form of rhetoric. The dominant spirit of our domestic and foreign policies has always been Puritan. This fact has gone largely unrecognized because we're taught that Puritanism was a New England phenomenon, centered particularly in Massachusetts, and we are given the impression that the phenomenon of Calvinism at some point evaporated. What is missing in this uncritical assessment is an awareness of the unqualified commitment the settlers of Virginia and other Southern states made to self-aggrandizement and entitlement. If Max Weber can be trusted, nothing is more fundamental to Puritanism than this unfiltered devotion to self, to Southern plantations and Northern corporations.

One reason our democracy doesn't function any better than it does is that it was never designed to be one. During our Revolutionary period, probably the most influential political thinker for White Americans was John Locke, possibly because he was notably gifted with "voice." His writing had what many an author strives for: a way of stating things that comes across as merely a careful exposition of common sense. In actuality, the things he wrote weren't as consistent as they may have seemed. The "good Locke" is echoed in the Declaration of Independence in his doctrine that there are natural human rights and that governments have a responsibility to protect them. The "bad Locke" is present in our Constitution in the special privileges it provides holders of property—and particularly in its way of providing extra electoral rights for those who claimed enslaved Africans and their children as property.

Locke, in his alignment with *vacuum domicilium*, put forth the notion that land and resources (water, trees, and so on) become privately owned when an individual mixes personal toil with them to make them "productive." Evidently for him—and Jefferson, Madison, Monroe, Jackson, and Polk—capturing

Africans or securing title to captured Blacks was "toil," and the stalwart Whites who performed this toil of securing title thereby turned the captured humans into "property." Locke—good at evasion when it suited convenience—did not, of course, put it quite that baldly. What he did was pen a defense of slavery regarding captives who had performed as enemies in war. Conveniently, the cause of these hostile captives had to be interpreted as an unjust one—otherwise his writing would have been a flagrant defense of sheer kidnapping.

Notice how profitable Locke's case for keeping captives as property was in real-world transactions. Anyone who captured someone in war was almost certain to say the captive had acted improperly. Any victor could conveniently "spare the life of the evildoer," as he might put it, and keep that person as perpetual property. And since, in many societies historically and later under colonial law, the children of slaves were slaves as well, what a windfall for any victor in a war—and what a fine incentive for waging one.

Locke's defenders might say this is just a slight exception to his overall doctrine that all humans are equal. Yet could any captive regard this exception to equality as "slight"?

I use the term *convenient* in describing Locke's maneuvers. Defenders of Locke generally allow that it was he who wrote in an official statement of provisions for Carolina: "Every freeman of Carolina shall have absolute power and authority over his negro slaves." Some hasten to add, however, that in so writing, Locke was acting as a lawyer for Charles II. "If he wished to be a lawyer for Charles," they might say, "he *had* to write such things." How quickly such defenders glide past that *if*. When the same argument was used by those on trial at Nuremberg after the Second World War ("I was only following orders!"), the accused often had as exculpation that they would have been executed had they not obeyed. Locke did not have that excuse. What he had was a well-practiced sense of how to play hide-and-seek. (One of the most desperate defenses I've read of Locke actually argues that we should remember Locke didn't invent slavery. This is like saying "Even if one of the Borgias did poison his enemies, we should remember that he didn't invent poison; use of it was a longstanding remedy for disputes between neighbors.")

Undisputed is the assertion that Locke, Secretary to the Lords Proprietors of Carolina, was a major holder of stock in the Royal African Company, which was established in 1660 to trade for gold but quickly shifted its focus to trade in captive Africans. Defenders can ask, "What could Locke do? King Charles awarded him the stock—Locke was a victim here!" (In the next century, this well-worn baton was passed onto Jefferson, who actively sought credit for being a slaveholder, regarding himself as a "victim" of King George.)

Unfortunately, in the confused state of American politics, Locke is a hero to people of both parties. People on the right praise him for his advocacy of limited government, claiming it means enterprising, up-by-their-bootstraps agents can act free from the heavy hand of government regulation. People on the left praise Locke for his championship of the equality of human rights but have little awareness of how tortured it was by adherence to Puritan theology.

Aristotle's approach to slavery was less convoluted than Locke's. In his *Politics*, Aristotle shamelessly taught that free Greek women were rationally inferior to free Greek men, but that barbarians were "natural slaves," rationally inferior to Greek women. What this shows, I think, is that even as critical a thinker as Aristotle could make the profound mistake of confusing tribal custom with "natural order." Privilege and presumed entitlement were a trap.

Both Aristotle and Locke fell short, then, in the matter of universal human rights—Aristotle explicitly and Locke surreptitiously. Aristotle, however, provides a clean contrast to Locke with regard to Locke's second claim to fame—his doctrine of limited government.

Let me try to schematize.

1) We "Locke'd-in" Americans think the human is a creature who may find it convenient to create a government. If the government does not satisfy, we may dismiss it—and may even decide, as republicans, to have no government at all. Society too, then, is a kind of secondary thing. We already-complete individuals may or may not find it beneficial to belong to a society.

By contrast, Aristotle taught that we humans are social animals. Not only do we create society but—equally important—we are created by it. Our only means of becoming genuinely human is to be profoundly socialized. The alternatives are that one be "a beast or a god."

2) Locke'd-in Americans think relations with other humans are *extrinsic.* One seeks relations with other humans for convenience; we can take them or leave them. (Locke died a bachelor, and it seems commonplace for a majority of his philosophical progeny today to be unable to sustain a lifetime partnership with a spouse.)

Aristotle taught, as did Confucius, that relations with others are *constitutive.* That is to say, it is out of respectful friendship with good people that one achieves a wholesome moral identity.

3) Locke'd-in Americans think *property* is the human good. We live to acquire. The richest are the best, the most blessed, the most admirable. (Both Locke's parents were Puritans.) If humans

can acquire property, no one can fault them for doing so. A human is "completed" by *having.*

Aristotle taught that the good of humans is in *doing*. We are made happy by reason of actualizing our best potentialities. The way for humans to become happy is by exercising their potentialities benevolently. It may be difficult to locate this teaching in a given phrase, but what we have is the basis for his belief that virtue is its own reward.

4) For Locke'd-in Americans, "important" property tends to be private. What is important to me is what I personally own. What is not mine but belongs to the public is alienated from me. (Infrastructure and public healthcare belong in this latter category, and are regarded as "externalities" by Milton Friedman and other faithful Lockean economists.)

Aristotle taught that even private property should be dedicated to public purposes. Nothing is more gratifying to enlightened persons than the gifts they have made to others. This is why it can make sense even to die for one's country.

5) According to the Locke'd-in, the function of the government is to safeguard human rights, particularly the right to private property.

Aristotle taught that the function of government is to serve the common good, creating conditions (peace, health, education, and so on) for the exercise of friendship and so provide for wholesome living and celebration.

6) The Locke'd-in often believe that achievement of domestic and foreign national security for Americans is the sum of the government's warrant; they believe that when government goes beyond this, it oversteps. (Devotion to the military-industrial complex is a natural outcome of this conviction.)

Aristotle would certainly have agreed with Thomas Hobbes about the primacy of survival (*Primum vivere*), but he thought achievement of national security was only the *first* of government's responsibilities. Security was not the end but the beginning.

7) Lockeans often believe that the good life is to be free from the performance of civic duty, secure in one's possessions.

Aristotle granted that an ability to contemplate—as presumably the gods do—is to achieve the pinnacle of happiness. Short of that though, he seems to have thought the good life is to be free *for* the performance of civic duty—and the goodness of life is, in fact, to be found in the actual performing.

8) Most Lockeans believe it is a mark of proper government to leave the good person alone.

Aristotle taught it is a mark of enlightened government to be involved in promoting well-informed citizens—citizens educated to think critically and to be ungrudging custodians of the common good.

Today, both major political parties seem to accept Locke's platitudes uncritically. More precisely, I'm concerned—as were historical scholars Charles Beard, Mary Ritter Beard, and Howard Zinn—that people who regard themselves as liberal Democrats largely fail to notice how much Locke's niceties about human rights are diluted by his doctrines on how property rights to own humans are acquired.

Locke's agreement with *vacuum domicilium* was wonderfully convenient. Think how odd it is that Jefferson could regard Napoleon as the title holder to the vast lands that became the Louisiana Purchase. Once the purchase was transacted, as American settlers moved westward, they could and did regard any Indigenous Americans encountered as trespassers on United

States soil. And American settlers treated these "trespassers" as they deserved to be treated—they could either submit or be killed.

As for money, Locke declared one can acquire as much of it as one can, provided doing so does not infringe on the rights of others. To me, this is Lockean convenience at its most obvious. He seems to believe a person's accumulation of immense amounts of money can be a private matter. The untruth of such a claim is manifest. In actuality, we live in a world that routinely does not meet the basic needs of all. Therefore, the distribution of goods is achieved through an ongoing auction in which the rich routinely outbid the poor. Will only some have potable drinking water? We know who these will be. (Not Flint, Michigan.) Will only some have benign police forces? We know who these will be. (Think Beverly Hills.) Will there be excellent medical care for some, while others live in desperation? Of course. Will schools be better equipped in some neighborhoods than others? We already know where those neighborhoods are. And Democratic politicians tend to be as determined to die millionaires as any Republicans they denounce, both sides feeding at the same trough.

Yeats could have been talking about us when he wrote: "The best lack all conviction, while the worst / Are full of passionate intensity."

If, though, we are to emerge with anything worth saving amid the unparalleled challenges that face us, it is through government that we will do so. If we are to have a future, we will have one only with government sponsorship of major efforts to control or manage global warming, rising sea levels, and magnifying storms. We will have to work together to create compensations for loss of arable land. We will need to undertake a variety of programs to provide water where there is increasing aridity and a growing number of people.

The Enlightenment casts a ray of hope across the foundations of our nation. Through a summoning of our social impulses, we humans can do these things. We can do them by tapping into the human genius for cooperative activity that can lead to a government that entertains assignments greater than the paltry prescriptions of the Lockeans, Jeffersonians, and Friedmans of our past and present.

Chapter 11

Milton Friedman and the Pursuit of Profit

WHILE A REDUCED and distorted version of Adam Smith had become a standard feature of American thinking well before Milton Friedman's time, it took Friedman to endow it with that character of the "one true faith" that is a contemporary trademark of American Puritanism. If the business schools of America today teach a bottom-line, scorched-earth approach to economics, Friedman more than anyone else can claim the credit. The late Friedman, of the University of Chicago and later of the Hoover Institute at Stanford University, is regarded by some as America's foremost expositor of the "conservative philosophy." His position is called "conservative" because its controlling concern is to clarify and promote the laissez-faire principle Smith set forth, which says government should leave the marketplace free from its intrusion. Friedman himself then is called a "conservative" for being dedicated to conserving this principle—a principle he thought modern-day liberals had a terrible habit of diluting or betraying.

One can catch the flavor of his conservatism in an opinion piece published by *The New York Times Magazine* in 1970.

Chastising some business executives, Friedman wrote:

> The businessmen believe that they are defending free enterprise when they declaim that business is not concerned "merely" with profit but also with promoting desirable "social" ends; that business has a "social conscience" and takes seriously its responsibilities for providing employment, eliminating discrimination, avoiding pollution and whatever else may be the catchwords of the contemporary crop of reformers. In fact they are—or would be if they or anyone else took them seriously—preaching pure and unadulterated socialism. Businessmen who talk this way are unwitting puppets of the intellectual forces that have been undermining the basis of a free society these past decades.[1]

What Friedman objects to is the doubleness of vision that a concern for social goals can introduce into the thinking and behavior of a businessperson. Sitting in a classroom a couple hundred yards from the Hoover Institute during Friedman's residence there, I remember hearing a very orthodox Friedmanite economist warn us that in the event we become businesspersons, we must not sentimentalize the workforce; rather we must remember always to regard "it" purely and simply as a "factor of production." If, by laying off half our employees, we can increase the profits of our company, it is our duty to the

[1] Milton Friedman, "A Friedman doctrine—The Social Responsibility of Business Is to Increase Its Profits," *The New York Times Magazine*, September 13, 1970. My own comments are a somewhat revised version of comments I first made in my earlier book, *Why the Center Can't Hold: A Diagnosis of Puritanized America* (punctum books, 2016).

stockholders to do so. *Not* to do so would be an act of theft—of stealing from the stockholders. (And if there are no stockholders? No doubt then, it would be a wasteful crime against ourselves to spend more money on the workforce than we had to.)

I remember thinking as I sat there how sublimely surgical this thinking was. Two thousand five hundred years of Hellenic and Judaic humanism out the window with a single toss! In its place: "Always treat a fellow human being as a mere means; never allow fellow feeling or a regard for the individual to confuse your thinking; realize always that for all practical purposes the employee is an instrument." Friedman did not say those exact words, but he would have agreed with them.

Years later I was in a college workroom running off handouts for a class when an economics instructor came upon me. He asked what I was up to. When I said I was preparing for an ethics class, he replied that ethics was good for Sunday afternoons but that the rest of the week was controlled by the discipline he taught.

I loved him for his unvarnished arrogance.

I reflected how completely Friedman's sense of what was important and what was expendable had prevailed—with this instructor at least. But probably, in truth, it had prevailed as well with many of the students I hoped I was encouraging to live ethically. To be sure, Friedman likely never felt liberal "reformers" and their "catchwords" genuinely constituted a worthy Sunday afternoon diversion. Rather he was afraid people were genuinely paying attention to these "catchwords" and "false prophets"—else why the essay in *The New York Times Magazine*?

At the end of the short excerpt above, he says that those who listen to such reformers are "unwitting puppets of the intellectual forces that have been undermining the basis of a free society these past decades." Anyone who attempts to introduce into the workplace an ethics distinct from the discipline of the free market itself, he's saying, is not far from committing treason. He argues, in fact, that his free-market economics is the *only* actual ethics.

He would have it that those attempting reforms betray our free way of life—betray, presumably, what we send our young overseas to die for. The executions of "traitors" Julius and Ethel Rosenberg in 1953 weren't yet a distant memory as Friedman was writing. They were context, and Friedman's use of the phrase "intellectual forces" was meant to convey that people like the Rosenbergs and fellow travelers were the agents of forces willing to ruin us all. Friedman took his political opposition seriously, and for this reason—notwithstanding his trademark smile—we should take him seriously. Today when we hear echoes of his rhetoric on Fox News and talk radio and in the rants of Trump, we shouldn't shrug off these voices as mere chatter. They express the set of beliefs in which our highly materialist, self-referential, and narcissistic Puritan tradition is rooted.

In exposition of this tradition, a few paragraphs later in his opinion piece comes Friedman's effort to elucidate how the profit motive provides the one and only responsibility of a corporate executive in a typical business:

> In a free-enterprise, private-property system, a corporate executive is an employee of the owners of the business. He has direct responsibility to his employers. That responsibility is to conduct the business in accordance with their desires, which generally will be to make as much money as possible while conforming to the basic rules of the society, both those embodied in law and those embodied in ethical custom.

He continues:

> Of course, in some cases his employers may have a different objective. A group of persons might establish a corporation for an eleemosynary purpose—for example, a hospital or school. The manager of such a corporation will not have money profit as his objective but the rendering of certain services.

Friedman's cards are face up. His candor shouldn't disarm us. It should draw us to examine critically the only game he takes seriously.

Doesn't it seem a bit quaint, charming even, to describe a CEO as an employee of the stockholders? It suggests that CEOs are like assembly line workers, spending their days in carrying out the orders of their managers—or that CEOs relate to their stockholders the way a maid relates to the mistress of the house, curtseying at "Today, Hilda, I want you to dust the windowsills."

The quaintness resides in Friedman's implication that the CEO spends his or her day consulting with stockholders in order to do their bidding. Friedman might counter, "The CEO doesn't have to consult his employers; he already knows what they want. They want—as I said—as much money from their stocks as they can get." Still, though, one wonders about his accuracy. In the name of schematization, doesn't Friedman oversimplify?

Corporations get started for a purpose. There is a rationale for the corporation before there's any rationale for stockholders. Unless one is a mere scam artist, it is improbable that one begins by saying, "I'd like to make a lot of money; I think I'll get a lot of people to invest in a corporation." Doesn't what typically happens resemble more closely the way, for instance, someone like civil engineer Theodore Judah—living in the 1850s in the promising but remarkably isolated new state of California—comes along and says, "There ought to be a transcontinental railroad"? Only when that has been said does Judah venture forth in search of investors. (For some of the seminal ideas here, I'm indebted to a brilliant article by Christopher Stone, "Why Shouldn't Corporations Be Socially Responsible?")

Maybe Friedman would counter, "Well, the motivations of the investors can be very different from the motivations of the visionary. The visionary may not care about money, but the investors do." Judah died at thirty-seven of yellow fever, as chief engineer of the Central Pacific Railroad. Suppose, though, he had lived to become—as he had hoped—the CEO. Would his motivation have been simply to make money for his investors, or would it have been to construct and run an excellent railroad? From what we know of the man, quite probably his original vision—with its quest for quality—would have carried through into his career as CEO. When a business has quality in its goods and services, it is highly probable that its inspiration was to provide some kind of social benefit. Its roots are altruistic even if profit is one of the fruits it produces.

Would Judah have been guilty of "fraudulent" practice had he survived and then gone on to seek the best steel, lay out the best track on the best route, and then provide the finest service he could? Had he been so accused, he likely would have responded, "How can this be fraudulent?" He would say the investors must have known the purpose of the man and of the project they'd invested in.

Cynics may reply this didn't happen—not in Judah's case.

But Judah puts us in mind of how money—in the best cases—begins as the facilitator and then becomes the maintaining consequence of well-considered projects. The project continues to be the goal. Money is an instrument.

It is Friedman's reduction of everything to money that is uncritical. This disqualifies him as an authentic disciple of Adam Smith. Smith understood the function of money was to facilitate life. He rejected mercantilism—that is, government control of minute decision-making—and wished to free up and protect the small-scale transactions motivated by private demand that were of concern to him. His thinking and the later thinking of John Maynard Keynes, a champion of infrastructure, were, I believe, closer than people realize. Friedman seems, on the other hand, to take it for granted that people invest only to make money. He doesn't take seriously the possibility they invest in things they believe in—that they invest to make particular things happen. When they do, he seems inclined to think, "Now we're no longer actually talking about business; we're talking about hobbies."

To structure things thus—to turn things around the way he does—Friedman converts his model of what business is into a Procrustean bed. To save his model, he's willing to lop off any areas of human endeavor and human interest that the model can't neatly accommodate, implying that there is something going on that isn't at all what business is about. To consider an example: It doesn't seem that Pope Julius II invested in Michelangelo in order to make money. Nor does it seem that Michelangelo's aesthetic sense was guided by some compulsion to see that Julius II got a maximum return of monetary profit from the ceiling he was painting. Friedman's objection to anyone's invoking this particular kind of enterprise is predictable: "I'm not talking about *that*!" He might add, "The word *profit* doesn't even come up in such transactions!" Exactly.

And that position should alert one to be thoughtful about how long a list one could generate of other *that*s about which it never occurred to Friedman to talk. True, he spoke of the business executive as making "as much money as possible while conforming to the basic rules of the society, both those embodied in law and those embodied in ethical custom." This seems like a good and necessary concession—surely one providing space for Adam Smith's reflections on how free markets would operate to produce the greatest good. But one must wonder what weight Friedman allows this concession. After all, he is arguing that business motivated by profit is sufficiently disciplined by the free market itself. The role of government, and presumably of government law, is to leave the market alone.

He says, for instance, that the business executive should not worry about pollution of our habitat. Friedman's position is either that there is no ethical custom of taking care of the environment or—if he recognizes that as indefensible—that making protection of the environment normative is an error in ethics since it would lead to "pure and unadulterated socialism" and draw one into the ranks of "unwitting puppets." So much, then, for ethical custom. And no wonder we ignore climate change.

We arrive here at one of Friedman's more troubled moments. Suppose our government grants corporate status to some group of businesspersons. This act confers a fictitious personhood, and the government endows this "person" with certain privileges so it can achieve goals that the government approves as conducive to the general welfare. Yet government should not, according to Friedman, intrude upon the operations of the very things that its law has birthed into existence. It must become an absentee parent.

Friedman's intentions about conforming to law and ethical custom are so vague as to dwindle to a null set, because, taken seriously, they work against Friedman's program. He professes to address the real world. In the real world, it is no great chore for corporations to buy the government and then have it legalize whatever the corporations are up to. This is what lobbyists do. (Graceless genuflections of John Roberts and other members of the Supreme Court have made it easy for people of wealth to own the government. This was the goal of *Citizens United.* It was not an accident, nor would it seem to be anything Friedman would have deplored.)

I don't mean to assign to Friedman any motivations he didn't publicly endorse. The title of his *New York Times Magazine* article is "The Social Responsibility of Business Is to Increase Its Profits." But as a matter of fact, corporations don't incorporate merely to increase their profits; it's been the custom for governments to provide them privileges so they can build bridges, mine coal, feed the hungry, cure the sick.

One should, then, think in terms of contractual obligations. CEOs ought not simply look over their shoulders to their stockholder-bosses but ought to look back further to the particular purposes for which their corporations were authorized. Yet in Friedman's treatment of his model, the dimension of regulation—that is, concern and guidance and requirements for corporations—would dissolve the moment a corporation comes into existence. At that moment, the purpose of the corporation becomes internalized.

This is not good even for the corporation. Enron, on the eve of its implosion, was such an internalized corporation. One cannot legitimately assign to Adam Smith an endorsement of a doctrine that it is best to have unfettered corporations. For Smith, the pursuit of the common good was the motive for curtailing mercantilism. Smith's famous invocation of an "invisible hand" was motivated by this commitment to nurturing the common good. This commitment incites in Smith a cautionary concern: Just as he didn't want government to be in control of all business decisions, so too he didn't want business mergers (trusts) to have that kind of power.

Smith regarded with apprehension the special privileges that merchant guilds, cartels, and other such business associations tend to assign to themselves. He thought such bodies could be as much in restraint of trade as governmental mercantilism could. In his view, monopoly was monopoly—whether stemming from governmental mercantilism or "big business," it restricted access to enterprise by outsiders and harmed the public interest. Think here of the deceitful united front that oil companies have joined to deny the evidence of a global threat to which they gave earlier testimony. (A video by Shell Oil once testified to the danger of global warming from carbon dioxide. Presumably that video was intended for the classroom. It has all but disappeared.)

To repeat, it would not have occurred to Smith to claim as Friedman does that once a set of privileges is in place, government has no responsibility for the impact the exercise of those privileges has on the opportunities of those who are granted no participation in them. While Friedman is concerned for the freedom of fictitious persons called corporations, Smith is concerned for the freedom of three-dimensional human beings.

We see that Friedman's corporation is born of his capacity to abstract. It is conceived as unconnected by real-world relations to things outside itself. It is not beholden to the fire department or the police department for its safety. It is not beholden to some system of public water. It doesn't use roads. It makes no demands for protection against foreign invasion. It owes nothing to the ends for which it was instituted. The educational system that has provided it with qualified workers is not entitled to its support. Only on such assumptions is Friedman able to find that the corporation is without genuine responsibility other than to increase the profits of its stockholders.

Friedman's corporation is a fantasy. He is serious in claiming benefits and opportunities for business, but if, when he was among us, one had asked him about its corresponding responsibilities, he'd have wandered off into a denunciation of socialism. It is especially noteworthy how, in the first of the excerpts above, Friedman dismisses "avoiding pollution" as among the "catchwords of the contemporary crop of reformers." Grown-up businessmen, he seems to say, don't concern themselves with avoiding pollution; the environment is an "externality." In tribute to Friedman and the tradition for which he speaks, one can't help but be impressed at how thoroughly this lesson has become embedded. Whether Friedman teaches this lesson of indifference because it's simply the rule in play or whether it's the rule in play because he and like-minded economists teach it is, I suppose, hard to measure. My hunch is that the dispensation from caring for the environment granted here by Friedman and his "conservative" predecessors and current disciples contributes hugely to our current irresponsibility in the face of danger. If pollution is not the concern of business, neither, then, is global warming.

Consider how in 2010 incalculable amounts of oil poured into the northern waters of the Gulf of Mexico from a pipe in the water below the *Deepwater Horizon*, where BP (formerly British Petroleum) was drilling for oil. At that time, a guest expert on *PBS News Hour* cautioned watchers against "hysterical reactions." From the general tendency of his remarks, it was clear that it would *not* be a "hysterical reaction" to be deeply concerned about the future of BP. All adult persons know BP is important and must, at all costs, be preserved. "Hysterical" would be worrying so much about the Gulf of Mexico as to put the future of BP in jeopardy.

To recast this expert's sentiment in the vocabulary of Friedman: It was never the function of BP to "protect the Gulf"—a "catchword" if ever there was one. The function of BP was—and is—to deliver oil. Yet even that is perhaps too romantic a description. The real function of BP has always been to deliver money to stockholders. That is why BP is right not only to drill for oil but right to drill for it in as quick and cost-cutting a manner as it can get away with. (The PBS expert must feel happy; we seem to have recovered from hysteria, and the future of offshore drilling remains bright.) One can hear Friedman muttering from his grave, "That's just the way it ought to be. If you're looking for someone to blame, that's a sign of your immaturity—for this is the lifestyle you've been lucky enough to get born into. To think otherwise—to load BP with concerns about the environment and attempt to shame it for not living up to those concerns—is to be among the 'unwitting puppets of the intellectual forces that have been undermining the basis of a free society.'"

As we've seen, Friedman does acknowledge this isn't the whole picture. He speaks of "eleemosynary" pursuits. That the adjective is rare enough to function as a stumper in spelling bees is instructive. Friedman suggests by the term that we pass at this point outside the purview of serious business and into "charitable" works. We've already noted the arrangement between Pope Julius II and Michelangelo as one example, and Friedman offers hospitals and schools as others. He acknowledges that in these enterprises the purpose may be other than profit maximization. He makes a conscientious concession that not everything fits his model.

Notice the attributes of what we consider here. There is an undeniable need for both treatment and maintenance of our health and the distribution of knowledge in our society. Hospitals and schools meet the need. Both require funds to operate, but the purpose of neither is properly reducible to the making of a profit, though many a so-called conservative would like to make it so. What Friedman is not about to consider is that these business enterprises—hospitals like St. Jude's (no charge) and tuition-free state universities—could actually be excellent models for reconfiguring our thinking on business in general.

Let's say that to be considered legitimate any institution must be founded to fulfill some wholesome function. Well and good. Yet income in excess of daily expenses is necessary if such institutions are to survive. Again, well and good. The surplus, or "profit" if you want to call it that, goes then toward maintenance and renovation—or public funding through taxes could support the funds for such maintenance. Money is instrumental and necessary, but it's not the foundational goal. (At this point, the Friedmanites recognize socialism and are shouting "Treason!" For just this reason, they are smacking their lips doing all they can to convert schools, prisons, hospitals, and all forms of healthcare into for-profit enterprises. Like the dutiful Puritans they are, they want to rationalize all operations of American life by disengaging them from sentimental trafficking in notions of the common good. Only in that way can the managers establish their bona fides as good Puritans. That many call this monetary form of advocacy "Christianity" is a truly impressive acrobatic achievement.)

A decade ago, I complimented a doctor laboring in the hectic emergency room of a crowded public hospital for bringing my son back from a coma that could have ended in death. I added to my remark, thinking perhaps like Friedman, "But you don't get paid adequately for the work you do!" She replied, "Oh, I get paid enough." I persisted: "How so?" She said, "What I get paid permits me to do the job I live for."

The doctor declared a truth. The contented doctor (who may today be an endangered species) does not go into medicine to make a lot of money but rather to cure people. The happy lawyer (if there is such today) goes into law to protect rights, curb wrongdoing, secure justice. The happy teacher is sustained by the experience of helping students learn.

Aristotle, who would have gotten an F in any economics course taught by Friedman, endorses what my son's doctor said. Insisting that "money is means," he adds (I paraphrase): "One who lives for money as one's goal lives the life of a slave, for there's no natural completion for his pursuit." He elaborates: "Happiness is in doing—it is achieved in action performed in accord with virtue." My son's doctor exemplified this point. (In my experience, many an ethics student, feeling things have gotten a bit mystical with Aristotle, wants at this point to return to "money as one's goal." Once he has declared that money is "pure instrumentality," though, Aristotle is not about to back down.)

Friedman, in his trademark way of simplifying, ends with a bifurcated and unintelligible world. This is the world in which we now live. Friedman would have the father of children feel affection for those children and concern that they live well in times to come; yet he would have that same father go off to work and labor each day simply to increase the return to stockholders without calculating the impact his corporation might have on the society his children are to live in. Friedman might respond, "Well, that's just the way people behave," but he complains rather that they don't behave this way *enough*. He wants precisely this behavior. He did his best to dictate it to the business schools and legal firms of the land and thought people were behaving unethically when they didn't practice it.

More insistently than Adam Smith, Friedman was a missionary. It turns out that Friedman wasn't just about money after all; he was about his *doctrine* about money. And this is what makes him a Puritan. Not only is the world that's generated by his thinking an alienated one, but he insists we enter this world as a destiny—as our ticket into adulthood. To our misfortune, we often do. There is nothing in nature that requires a businessperson to shed the responsibilities of a citizen when becoming CEO. Somehow, though, Friedman feels businesspeople can and should be forced to do this—should suspend their humanity on the grounds that they are owned by stockholders. (As Max Weber claims, capitalism makes people prisoners of their possessions.)

While Friedman speaks of the restraints of law, there is nothing in his system to keep corporations from using their money to buy lawmakers and insist that they provide corporations a maximum of privileges and a minimum of responsibilities. Some people were surprised when Friedman offered himself as a mentor to General Pinochet, the Chilean dictator who replaced President Salvador Allende following Allende's death by suicide after he was ousted (with assistance from our CIA). No one should have been surprised at what Friedman pitched. Fully provided for in his conception of freedom is the freedom of a dictator to control all the laws and be accountable to none of them. He had already done what he could to secure that kind of freedom for American corporations. In the case of Chile, he intensified and "purified" the kind of free enterprise the US was diligently mandating for Latin America.

To be fair though, in asking us to join his world, Friedman—as I've tried to show—plays to the most deeply entrenched tendencies of our history and culture. This is why "liberal" Democrats are such lame critics of "hardline" Republicans. The Democrats emerge from the same history and culture and have much the same values and delusions. Routinely, when it has been a matter of actual practice, our Puritan heritage has trumped the shining parts of our Enlightenment heritage. When we want to feel noble about ourselves, we put on the rhetoric of the Enlightenment, but when it is time to act, we regress to the behavior of our forefathers that is more fundamental to our thinking: We go Mystic Fort.

Chapter 12

The Failure of Broadcast Journalism

JOURNALISM IN THE UNITED STATES underwent a profound transformation during the twentieth century, gradually shifting from a public service rooted in journalistic integrity to a business enterprise increasingly influenced by the profit motive and corporate interests.

Before electronic media, the news arrived in print. Almost everyone in a neighborhood subscribed to at least one newspaper. Subscriptions provided journalism with security—reliable income. This printed news, of course, carried ads, but no reader had to give more time to them than they wished. In the golden days of print journalism, most ads likely found their way to the wastebasket unread.

However, in the latter half of the century, news organizations began moving away from subscription-based revenue to an advertising-centric model. This transition gained momentum after World War II as radio and then television became the dominant medium. While the ad-centric model made news more accessible to the public by lowering or eliminating direct costs, it also introduced a significant downside: Advertisers became a powerful, often invisible influence over news content.

Even as the popularity of radio and TV news broadened, some naysayers believed that TV massaged us into a too-compliant mindset, including when watching the news. As Vance Packard argued in 1957 in *The Hidden Persuaders*, many persuasions and dissuasions are served up at moments our guard is down and we have little awareness of ingesting uncritical assumptions and unwholesome pitches.

More recently, in his 2016 book *The Attention Merchants*, Tim Wu underscored revered WWII correspondent Edward R. Murrow's resolute disappointment in the failure of the network executives—chief among them his onetime promoter, William Paley at CBS—for consciously abandoning the opportunity to use television as a champion for keeping the American people informed. Wu suggests that Paley's visits to the hospital during Murrow's final illness were an attempt to seek absolution for trivializing TV—a pardon Wu believes Murrow never granted.

When the controlling goal of network news became attracting the largest possible audience for advertisements, Murrow knew that the enterprise he thought he and Paley were creating together was being abandoned. Since then, network executives at ABC and, less completely, CBS have often acted irresponsibly, and they have done so knowingly. They have promoted a rosy outlook even when the reality has been blood and guts, presenting "if it bleeds, it leads" stories while insisting that the moorings of American society are sound. Think, for instance, of Bush Jr.'s irresponsible "Mission Accomplished." The deeply negative aspects of American domestic society and the deceits, militarism, and inhumanity of much in American foreign policy have not been denied so much as ignored.

Our broadcast media have often pretended to possess evidence they don't have. Just as tellingly, they have faltered when real evidence has been placed before them. Deep in mendacity, they have been awkward in how they feature brave whistleblowers like Daniel Ellsberg, Chelsea Manning, and Edward Snowden. Consider as well how the networks disgraced themselves by denying Julian Assange the defense he deserved. In domestic matters—with worldwide implications—our media have been slow to take a hard look at the protections and privileges granted Monsanto as it instituted monoculture wherever it could. The misleading premise seems to have been: "If Monsanto and Roundup were bad for America and the world, we'd never have let them become so popular." In other words, products are innocent until proven guilty.

Similarly, when DuPont began to understand that Teflon was biologically indigestible and harmful to mammals, including humans, one wonders why our news media did not pay more attention. The answer is simple: It would have been inconvenient to do so. Bad news of that sort sets a mood inconducive to commerce, which risks losing sponsors. Such news does not promote the mood for music-filled ads showing people discovering a magical life by utilizing some fresh wonder of consumer culture. TV has become the mercenary Murrow feared it would.

This mercenary mindset reinforces our domestic myths. TV provides a psychological climate favorable to scorched-earth capitalism by speaking routinely of the United States as "the land of the free and the home of the brave," with "liberty and justice for all." To speak otherwise is, in some quarters, regarded as treason: Knock not another man's sales, nor put a damper on the market. Worldwide, we have promoted the American brand—its products and its politics—by speaking of our politicians as standard-bearers for freedom worldwide. But if our leadership has truly led the free world, how free then is it? When our leadership blew parts of the Bikini Atoll to bits in 1954 and invited the Marshall Islanders to celebrate a great step toward world peace, were we unwittingly enacting a satire? (Perhaps merely one among many?)

Since the US Supreme Court's *Citizens United* ruling in 2010, it often feels as though our politicians are for sale to the highest bidder, their loyalty pledged to donors and lobbyists rather than the citizens they ostensibly serve. Money now shapes campaigns, legislation, and the framing of national priorities—and delivers a spectacle of political theater that television, in both news and advertising, amplifies, distracting viewers from the real performance and corruption of power.

Just as corporate money shapes politics, it shapes the way news is packaged for viewers. To the extent ads engage the news audience—whether for products or political aims—they redirect attention from the news. This seems obvious. What may be less obvious is that the managers of the evening news do not much *like* the news. They want David Muir to end a segment of the news with a sentence like "Stay tuned to hear of the sudden death of one of America's greatest actors" to hold the viewer's attention for another batch of ads. When David gets to the actor's name, it turns out this "greatest actor" stopped making films fifty years ago. But never mind—the actor's death has served its purpose.

WE AMERICANS HAVE a weak understanding of the world outside our borders. This is not surprising, since by the time our young reach the age of twenty, they have learned to expect little logical sequencing or media responsibility with regard to what is presented as news. The coherent world in which human rationality finds a context for situating facts within a larger perspective has slipped from reach. This lack of familiarity with logical process will be a theme in the next chapter. Note for now that today's young citizen is often mentally reconciled to a landscape of dots—without realizing the dots can be connected to reveal a pattern. (And while this obliviousness may seem to characterize students in a "disadvantaged" environment, that environment may be Harvard or Yale or Stanford.)

The consequences of this media negligence extend far beyond trivial distractions. They shape public understanding—or misunderstanding—of urgent issues, including climate change. On June 23, 1988, atmospheric physicist James Hansen testified before the US Senate Committee on Energy and Natural Resources about human activity as a cause of global warming, drawing upon the emerging worldwide consensus of climate scientists. It was evident to anyone paying attention that the forebodings that had emerged during the 1950s were being confirmed. Seven years later, climate researcher Benjamin Santer summed up the findings of the Intergovernmental Panel on Climate Change (IPCC) by writing that there was evidence of a discernible human influence on climate change.

What did the media news managers do with these statements? They deflected and reported instead on the contention of the oil executives that there was no significant news to report. Were they summoned by law to defend themselves, the managers would answer, "Wait a moment! We did exactly what responsible news agencies should do—we reported on the controversy!" The implication: A controversy is a controversy, and it is not the networks' business to investigate who's right and who's wrong; their job is simply to report it—and sell ad space.

The motivation of the oil executives and ad managers could not be more manifest nor more mendacious. The real controversy was not whether climate change was real but rather what to *do* about it. Toward the end of 2022, for instance, there were twenty or more extremely destructive tornadoes, along with fires and hurricanes of rare power, and strange earthquakes and floods; similar events have continued in the years since. Yet mainstream media and oil executives continue to downplay the evidence and withhold acknowledgment that anything beyond natural variance is affecting our climate. Two halfway alert TV journalists could have spent about fifteen minutes working together to become confident that the oil industry's testimony held no weight; the fact-checking would not have been difficult. Readily available sources provide an array of detailed articles on global warming to inform anyone interested in its reality.

Executives still don't admit, even to themselves, how much it has been the custom of their networks to use their news anchors robotically. Viewers' minds are wrenched from topic to topic with minimal connection or continuity. What remains are mere nuggets of climate news floating in puddles of commercials and trivia. It is as if Hansen's testimony in 1988 and Santer's in 1995 were never given and as if human history's most momentous story were not unfolding.

Psychologists and social scientists speak often these days of *cognitive dissonance*, a state in which someone holds two or more incompatible beliefs, or acts in ways that conflict with their values. The lack of harmony is denied, goes unnoticed, or is rationalized. Cognitive dissonance is everywhere. A person values honesty but cheats on their taxes because "everyone does it." Religious institutions know clerics molest followers and conceal the abuse to preserve authority. We know that gun violence kills children and yet rationalize it as "the price of freedom," refusing to consider any solution other than all-or-nothing. Compared with other high-income countries, America's rising suicide rate and high levels of addiction—to cocaine, heroin, methamphetamines, and more—suggest how unhappy many young people are and how they want to escape the configuration of reality that has come to register as "the way things are."

The networks, in their pursuit of ad revenue, go the extra mile in supporting our collective cognitive dissonance. They shovel misrepresentations, lies, and sheer nonsense into our brains rather than provide the information and facts we need if our conduct is to be responsible. Which celebrity is at odds with which is so much more important than finding out what meat eating and high-fructose corn syrup are doing to bring climate change and diabetes to our children. Empirically, we seem to care about our own kids as much as we cared about the kids of the Bikini Atoll whose homes we bombed into oblivion.

Currently, the atmosphere from which our network news arrives is rancid. When William Paley at CBS pursued nickels and dimes in the crannies of American commerce, and when David Sarnoff at NBC similarly refused to offend the priorities of hucksters who felt they knew the right mood for the sale of their merchandise, they tossed network integrity to the winds, and the news industry accepted the hollowed-out form it now embodies.

It should not surprise us that we who live in endless search for the signs of wealth that mark God's Elect are never satisfied by the prosperity of the gospel we've invented. The quest for money has no natural term. Why does it matter how each day's news is conceived? Considering the networks' indifference to the truth of the ads they run, one must wonder whether the news they present is treated with any greater care. If the ads can be made up, why not the news as well? Today there are many charges against the government for generating fake news, but didn't ABC by avoiding mention of climate change and CBS under Les Moonves's leadership shamefully perform in a way that helped establish a path for such reporting? Not least was Moonves's comment that the 2016 presidential campaign circus "may not be good for America, but it's good for CBS."

In fact, much of the network and cable news comes across as a gigantic conspiracy against the public. From the time of Hansen's and Santer's testimonies—and the consensus that virtually the entire IPCC endorsed—it has been clear that the accumulation of greenhouse gases reported by climatologists is being confirmed by manifest increases in heat and melting, storms and cold spells, fires and floods. Those who sponsor contrarian ads on the networks have access to the same data and the same reasoning as the IPCC, yet they pay networks to deceive the public, and the message they dictate is this: "Not only is this business about climate change controversial, but we regard network mention of global warming as hostile to us and to our attempts to sell products."

As I wrote this chapter, chunks of Kentucky were floating away into its sewers and rivers. The governor said he had never seen anything like it. Yet in such dreadful circumstances, many consider it unacceptable to play "the blame game." The proper speech seems to be "No one should try to draw a political lesson from this natural catastrophe!"

Yet for the last three decades, anthropogenic climate change has been as uncontested among climatologists as whether the earth is spherical. Our network news producers lie and have been lying for a third of a century now. They have lied by their silence, ignoring what scientists overwhelmingly agree on.

ABC and other outlets: "We never said the climate wasn't warming. We just said, quite accurately, that there is controversy."

Skeptic: "Yeah, but when something new of immense importance is happening, it is the function of journalists to find out what it is and to let the public know about it. That's why it's called *news*. You've ignored your duty."

Network spokesperson: "But our sponsors have been interviewing experts!"

Skeptic: "Have they though?"

Network spokesperson: "They've interviewed Freeman Dyson."

Skeptic: "Yes, we should give Dyson his due. He and his colleagues figured out how to obliterate huge swaths of the populations of Hiroshima and Nagasaki in less than sixty seconds, so one must agree with you and your sponsors: This makes him an outstanding expert on climate change and global warming."

Network spokesperson: "Okay then, there's no need to get nasty. Among our sponsors is Charles Koch, one of the richest men who has ever lived. That has to count for something!"

Skeptic: "Of course it must. For didn't Aristotle say, 'He who makes the pursuit of money the controlling goal of his life—and succeeds at it—must be the wisest of men'? Or wait—I got the quote a little off. What he actually said was that a man who makes the pursuit of money the goal of his life is among the most *un*wise of men—for his error is not a tactical but strategic one, for he misunderstands the instrumental character of money."

The general practice of mainstream news demonstrates contempt for the viewing audience, and the audience has become so used to it that it hardly notices. The networks also demonstrate a lack of respect for their own reporting staff, leaving their celebrity anchors to operate up to their eyebrows in the tainted waters of hucksterism. The *caveat emptor* premise on which ad executives operate is indifferent to humanity's future. As mentioned, they regularly presume a product is innocent until proven otherwise. They don't *really* believe this, but it's convenient to pretend they do. They leave it to the viewing audience to protect itself against groundless blandishments. If cognitive dissonance is running amok through the body politic today, think how it is pumped daily into the public domain by the TV networks and other media that seldom or never mention global warming. The miasma filling the brains of viewers parallels the brew of greenhouse gases collecting overhead.

None of this is new. In *The Attention Merchants*, Tim Wu dispassionately musters an impressive set of witnesses. His roster of critics of how we trivialize attention includes not only Edward R. Murrow but also Vance Packard, Herbert Marcuse, Marshall McLuhan, David Halberstam, Jerry Mander, and Walter Lippmann. Lippmann offered one of the most uncompromising indictments of the TV industry in 1959:

> There has been, in fact, an enormous conspiracy to deceive the public in order to sell profitable advertising to the sponsors. It involves not this individual or that, but the industry as a whole . . . there is something radically wrong

> with the fundamental national policy under which television operates. . . . There is no competition in television except among competitors trying to sell the attention of their audiences for profit. As a result, while television is supposed to be "free," it has in fact become the creature, the servant, and indeed the prostitute, of merchandizing.

Sounds like the concerns of Max Weber regarding the inability of our capitalists to exercise anything like a humane use of freedom, doesn't it? The "iron cage" is firmly locked.

Edward R. Murrow perhaps anticipated and outdid Lippmann in a lament he made a year before Lippmann's. Tim Wu (citing Erik Barnouw) paraphrases Murrow—that nothing on television should be too dark or challenging "for fear of contradicting the upbeat and optimistic commercial messages of television's sponsors." Wu makes the case for the corruption of TV in a prescient chronicle that is painstaking, convincing, and woefully predictive of present circumstances. Wu's conclusion is a worthy echo of Max Weber's iron cage clanging shut. He writes of the system that emerged: "Once that system existed, the logic of capitalism, though not decency, dictated it could be used for only one purpose: to make as much money as could be made."

Currently alive in the newsrooms of ABC and CBS—controlled as they have been by the pursuit of advertising revenue—there can't be much congratulation for enhancing the intellectual activity of the American people. When Scott Pelley attempted to take on Moonves's resulting wasteland, he was removed from the evening news.

Suppose we, the nominal beneficiaries of the evening news,

want to reflect on some matter just brought to light by David Muir. He appears to be an intelligent person who wants to deserve the free press protection that the First Amendment was intended to confer. The problem is, in the ranks above him, we arrive at banks of executives who regard the well-being of the viewing public as what Milton Friedman labeled an "externality." The executives are impatient with allowing the minds of viewers to germinate insights. They adhere to the "prosperity gospel"—that secularized religion Weber says Calvin invented and Milton Friedman eventually captured—in which the only sin is leaving money on the table. The CEOs are preoccupied with skimming our predicaments for opportunities of distress by which the enterprising among us can become billionaires. They're quite willing to impoverish the world and steal from their own children to do so. The ethical standard? Morality is whatever one can get away with—or can afford to pay a fine for.

EACH YEAR WE HEAR the networks reporting on wildfires raging through California. ABC and CBS praise, as they should, the heroic fire personnel who battle the blazes and sometimes die fighting them. The reporters commiserate with the stunned, nearly speechless people who have just watched their property burn to the ground. What we don't hear mentioned by the newscasters is that our cars and planes, powered by the internal combustion of petroleum, are agents of the devastation we see.

We seem to exercise our capacity for critical thinking best when we're watching or discussing sports. We cite data with conscientious specificity; we're alert to biases in commentary; we cultivate—amid remarkable feats of memory and judgment—

a shrewd sense of likely outcomes; and, if we want to have friends watch sporting events with us, we attempt to deal respectfully with objections to our preferences and predictions. It's worrisome that we seem incapable of maintaining this critical stance when it comes to public policy. We perform like stunned animals, or like helpless people in a hospital waiting room counting how many orange tiles there are on the floor.

People often claim they're not much influenced by advertising. If one takes that at face value, one must wonder why advertising firms get paid so well. Ask these "uninfluenced" people whether they can associate specific brands with the phrases, music, and emblems being used to advertise them. Most people do well at this game.

Advertising's governing strategy is to make the viewer feel inadequate: "How can I pass from the life of quiet desperation I lead and enter the life I see on the screen?" Ads manufacture need by stoking insecurity and offering a fantasy of transformation. As both critical theorist Herbert Marcuse and tech visionary Steve Jobs noted—one questioning desires, the other creating them—the expansion of human "needs" is elastic. If you can connect "ever-expanding needs" with "ever-expanding population," you're not far from grasping the scarcity taking hold in parts of the world.

Ads seldom enhance one's sense of well-being. They're meant to seed dissatisfaction, to convince you you're missing something. They're designed to kick into overdrive that desire French literary critic and philosopher René Girard described as *mimetic*, meaning imitative. Girard's premise is a simple one: We want what other people want. It is this impulse that drives people's desire to "keep up with the Joneses" or, to use a more contemporary example, the Kardashians. In making their pitch—"Others enjoy X; why shouldn't you?"—ads promote as the path to salvation the exercise of narcissistic preoccupation, or *dukkha*, the root of human misery according to the Buddha. This preoccupation stokes our tendency as a species to dwell in the grip of appetites. Jesus spoke of this danger in Matthew 12:45 when he warned that a man may no sooner clean his house than an evil spirit may return to it, bringing along other evil spirits worse than itself. Unlike Jesus, current leadership in America offers no warning; the promise of excess prevails.

Nuclear scientist Enrico Fermi seemed to think that we lack certifiable signals from consciousnesses outside our planet (that is, from "aliens") because any humanoid consciousness might be doomed by its afflictive desire for an indeterminate "more" and engineer its own extinction. This echoes Aristotle's teaching that the pursuit of money has no natural term. It can be argued that both men, but especially Fermi, also imply this about the pursuit of power. The lust for power lacks a natural limit but can go on until it destroys everything. We can do this with nuclear war, but we can also do it with climate change. There are serial killers among us, and there is a kind of panoramic totality in the manic minds of the killers. Their vision reaches out to everything that exists. They possess, and are possessed by, the program of the perfect capitalist.

When Girard wrote of mimetic desire, he regarded it as a source of envy and opposition—*the success of my neighbor makes me unhappy*—which can in turn lead to scapegoating and violence. Unlike some of his hate-saturated disciples, Girard did not celebrate mimetic desire as a means of crowd control. He saw it rather as leading to Jesus's crucifixion. The legacy of mimetic desire does not seem beneficial. Take just two behaviors that it passes on to young people today: First, it encourages the notion that a meal is not genuine and adequate unless it features meat. Second, it encourages the young not to worry about continuing the release of unaffordable carbon dioxide into the atmosphere. In the next chapter, we will consider what, if anything, education can do to counter these ruinous habits.

Chapter 13

What Will We Teach Our Children?

IN HIS STUDY REGARDING the good life, *Nicomachean Ethics*, Aristotle wasn't interested in teaching *about* ethics. He'd probably say few things were a greater waste of time. Rather, his goal was to help his students *live* ethically. In the tenth book of this work, the down-to-earth protégé of the unearthly Plato comes as near to echoing Plato's romanticism as in any document we have from him. Aristotle thought reality spoke in a special way to a person who lived ethically. The contemplative joy that arises from such a life is familiar to children, poets, and mystics. He says, in fact, that if anything survives one's death, it is the part of oneself that participates in the joy of contemplation.

Had Aristotle lived in later times and been able to read Thoreau's observation that most of us "lead lives of quiet desperation," he might have said, "This American has been eavesdropping on my conversations with students when we are out walking. While I have been charged with claiming that everything that happens to humans is in accord with reason, this fellow Thoreau has understood what I have actually taught—namely, that life would be much happier if that *were* the case."

Aristotle says that living to promote the common good fits

our nature but that most of us do so only occasionally. With the earthiness of Chaucer or the realism of Sancho Panza, he acknowledges that the meager use most of us make of our potentialities falls short of the excellence needed if one is to be consistently happy. He suggests that most of us prefer our sobriety and service to others be punctuated by flirtations with banality and compromise. This leads him, as it does Confucius, to emphasize the role that good parenting can play—how by words and example a parent can coax a child out of short-term narcissism and establish the child on a responsible course toward adulthood.

Yet even here, realism intervenes, and Aristotle all but takes back what he has said. While Plato says fatherhood should be anonymous, Aristotle makes fun of Plato, and though he insists that an attentive father can have an efficacy important for putting a child on a good path, he then enters a form of dialectic. Personal authority—father as teacher—is important, but this authority is most effective when it leads a child to respect good laws. Personal authority may in fact be resisted as burdensome, and the father's "tongue may be too light," whereas the law, when it is just, shines from the inside out and offers a compelling promise of order. Much is involved in a child's transition from family to society. The child is the parents' child, of course, but beyond that the child is a child of the state and will live a life of virtue and self-respect if they grow up to respect and promote the well-being of society. (If this makes Aristotle sound close to Jesus, who comes a few centuries later, Thomas Aquinas will eventually be a disciple of both in the Middle Ages.)

A corollary to a dedication to society is that society ought, therefore, to provide for the education of the child. In fact, a well-ordered state will take on the education of the young as a fundamental concern, since the endurance and improvement of the state will be a likely outcome of the younger generation's commitment to the common good. If a youngster is to grow up to live a life of civic virtue and thereby become happy, Aristotle says a program to achieve this should be underwritten and protected not merely by parental advice but by excellent laws. He sets this forth in the *Politics* as the rationale for his advocacy of a free liberal arts education. It is interesting that he does this in spite of the fact that Athens did *not* have a system of public education. His school, like Plato's before him, was a private enterprise, presumably paid for by parents.

Also, we must acknowledge another deficiency here. Aristotle seems to have viewed women as halfway between men, who ran things, and slaves, who carried out assignments originated by men. Women were intermediaries, but as caretakers of children, significant intermediaries. So he mentions the education of women as important but seems not to have made any provision for it. It's plausible that the services of women and slaves may have been a condition for the male's leisure to study and contemplate and govern. To make good use of Aristotle, then, it would seem helpful to extend his educational prescriptions beyond the boundaries he observed.

The term *liberal* seems to have multiple references as Aristotle uses it. An education is liberal if it is the education of one who is free from limitations imposed on women and slaves. Along those lines, something is liberal if it frees one from ignorance and a lack of opportunity.

Properly expanded, education as Aristotle proposes it ought to function as a bulwark against squandering the potentialities of the young. Here, though, a kind of tension emerges. Socrates was sentenced to drink poison for encouraging the young to address critically the ways of the older generation. Yet a genuine public education system would seem to be an enterprise sponsored by adults against whom the students were being prepared to hold their own. A well-educated student is one equipped to be critical of the status quo. Contrast this with the approach nowadays of, say, governors Greg Abbott (Texas) or Ron DeSantis (Florida), who seem to think students should be taught only such things as their parents decide will help them conform to the society the parents wish to pass on.

An Aristotelian education will be "liberal" then to the extent that it "liberates" the student to make excellent uses of freedom—both on their own behalf and on behalf of society. To repeat: The state must nurture an autonomy that will eventually empower the student to become its wholesome critic. (In his effort to provide for a meritocracy, Confucius had a similar concern.) The student is not thereby summoned to become unpatriotic but rather is summoned to an alert and transcendent patriotism. Consider, if you will, the example of Abraham Lincoln—born in the slave state of Kentucky—stepping forth to say, "A house divided against itself cannot stand." Lincoln was sustained and educated by a society that he would eventually seek to change.

A modern Aristotle would say it takes an excellent nation to guide the younger generation toward an excellent use of freedom. An excellent nation will plan and make provision for such an outcome. A nation that does not do this is not an excellent nation and is in jeopardy of decline, for its population will repeat its preferences and behaviors simply because they are familiar, regardless of whether they remain appropriate or were in fact ever appropriate.

A fresh question then arises: By what process does an excellent nation come about? Aristotle's *Politics*—intended to follow up on his introduction to ethics—is dedicated to defining the character of an excellent nation and describing how it can come to be.

It should not be a scandal to make the statement that just as an excellent nation is necessary if excellent citizens are to exist, so too excellent citizens are necessary if an excellent nation is to exist. Perhaps without the brilliant guidance of Pericles, there would have been no Athens to lead the Delian League, a confederacy of Greek city-states, from roughly 461 BCE to 429 BCE. Athens can be regarded as a symbiotic creation—the product of reciprocity between leader and polis. A virtuous people develop a constitution that conserves "the common good," and the common good preserves "a virtuous people."

Think about Darwin and step-by-step evolution. By the time the seventeen- or eighteen-year-old Macedonian Aristotle arrived at Athens in 367 or 366 BCE, institutions of excellence—the law, art, architecture, theater—had emerged through incremental developments. Aristotle no doubt felt that he and his contemporaries were fortunate innovators—frontiersmen emerging from a realm of rich potentialities. For all its risks to arrogance, there seems a notable measure of justification for such a way of thinking.

Regarding education, Aristotle makes two foundational claims: 1) Nothing is more important than that a nation have a clear idea of what it wishes the outcome of its educational program to be (Aristotle was a painstaking collector and student of the city-state constitutions of his time); and 2) Few things are more difficult than achieving consensus on this matter. Decisions as to curriculum are at once important and difficult—and may at times have more to do with the teaching of critical approaches and reasoning techniques than with the teaching of specific content.

Aristotle was innovative. As an instructor at the Academy, he seems to have expanded the curriculum beyond Plato's emphasis on math by adding an un-Platonic attention to "natural science," with particular emphasis on biology. There is reason to believe he encouraged the collection of biological specimens and modeled for students an engagement in empirical observation. This un-Platonic emphasis may have been partly why he wasn't selected to be the headmaster of the Academy after Plato's death.

The reputation that history eventually conceded to Aristotle was not an arbitrary matter. While Roman civilization seems to have been more inclined to be Neoplatonic than Aristotelian, nonetheless Latin, Arabic, and Jewish scholars continued to cite Aristotle with respect—Arabic commentators being especially important. Then came the Middle Ages and the Renaissance in Western Europe. Medievalists Albertus Magnus and Thomas Aquinas were Aristotelians. Later, during the Enlightenment, the new optimism about science brought a fresh interest in Aristotelian rationality. Then, as the Industrial Revolution intensified, not long before Max Weber's critique of capitalism, Pope Leo XIII used the encyclical *Rerum Novarum* (Of New Things; commonly translated as *Rights and Duties of Capital and Labor*) to reintroduce Aquinas and the philosophy of the common good at the very center of Catholic social teaching. Aristotle's way of thinking was once again quite relevant—and now Leo XIV, installed in 2025, has indicated he will continue along this path.

To the two claims noted earlier, Aristotle adds a third: The considerable difficulty of achieving consensus on the public education curriculum should not be taken as an excuse to *avoid* seeking such consensus. He seems to have thought a consensus was indispensable to the wholesomeness and stability of a society and to the happiness of its citizens. In the *Politics,* along with mathematics, ethics, politics, and biology, Aristotle sketches suggestions for including the study of reading, writing, drawing, gymnastics, and music.

If today we fall short in this business of public education, well, look to Aristotle. He never said it would be easy, and, as should be obvious, Athens only momentarily achieved the consensus he called for. Today, as in Aristotle's day, many issues are involved. When we speak of having a good educational system, whose education do we have in mind? Children of the upper class? Immigrant children? People who accept the Bible as normative? Everyone?

And what outcomes are we looking toward?

Should the knowledge the system aims to impart be confined to how to succeed in our culture as it is currently organized? Or should we be, as Aristotle (and Confucius) envisioned, raising students who will be fit to become the conscience and source of wholesome innovation for our society?

Should we be imposing merely a "vocational" education—one to make the young useful to the enterprises of the current older generation? Or should we be aiming at something more advanced?

Specifically, should this educational system aim mainly at how to get rich? Perhaps, confining our thoughts to Machiavelli's, we should say education has been achieved when the student has learned how to manipulate others and achieve an advantage over them. Milton Friedman and—I hazard from observation—

many of the business and legal departments of our great universities today would probably agree that helping students achieve a competitive edge in self-aggrandizement is what education should be about. A view supporting this is, as suggested, that all education should be "vocational in the broad sense," that learning is not an end in itself but rather a means toward the control and exploitation of others—or if not that, at least an end toward becoming an effective and dependable tool that others can use to control and exploit.

Alternatively, perhaps providing abundant information about the world in general should be the goal of education. But what would this approach say about, for instance, the much-exploited, nearly extinct, self-isolated tribal peoples in Central Africa today? Can such "primitive" people then be called "educated"? Yet it ought to feel strange to declare that people who have lived comfortably in rainforests for tens of thousands of years are "uneducated."

Still another option: Perhaps adept use of the scientific method—all the way to our own replacement by AI—is the ultimate and ideal success to be pursued in education. One sometimes gets the impression that logical positivists, Silicon Valley dwellers, and Elon Musk have some inclination in this direction.

For what it's worth, my impression is that we Americans, having deliberately destroyed children in many regions of the world, are now intent upon abandoning our own. It is not just that our legacy is one of miserable climate change. It is also a legacy of bad diet, poor healthcare, and incoherent educational systems. Students are taught that they live in a nation dedicated to the common good, where all citizens are created equal, and are also taught that they are specially favored by God if they can achieve an advantage over their neighbor.

Finally, contrary to all that, could it be—I ask sarcastically—that the desired outcome of education is to learn the two great "discoveries" offered by David Hume early in his career: 1) that we can never really know anything, and 2) that having acknowledged this, we should have the good sense to forget about it and conduct our daily affairs free from worry? While this approach offers an excuse for "party time in America," it really isn't much of a solution, for it appears—as Hume seems to have intended—to absolutize agnosticism or atheism.

Putting aside the question of which of these goals (or goals not listed here) should be pursued, we can acknowledge that one or a blend of them could be chosen, and that the results for society (and the resulting society itself) would be vastly different depending on the choice. In general, it would seem plausible that an educational system worth its salt seeks to provide some of the skills that must be mastered if the learner is to function effectively in their native society. While such mastery need not be the only goal or the organizing goal, it can be predicted that in actual practice it will be a prominent goal because, as we've noted, educational systems tend to serve the cultures and societies that sponsor them.

Let us now return to our discussion of the competing philosophies of our American heritage—Puritanism and the Enlightenment—and the implications for the American educational system.

Through the Fourth of July and Thanksgiving, American grammar school students are fervently taught the equality of citizens. As a corollary, they learn that one should dedicate oneself to serving the common good. ("Ask not what your country can do for you—ask what you can do for your country.") Here we have contemporary embodiments of Aristotelian teaching.

Simultaneously, students learn from their Puritan forefathers that they are specially endowed by God. This is not just theoretical: It means we are entitled both at home and abroad to make war and kill others in defense of our special privileges; to do less is to neglect our manifest destiny. We have taken our calling seriously. In fact, the continuing sign of our divine status is that we are able to achieve advantages at the expense of others.

But, as noted in earlier chapters, we face a complication. It was convenient during the run-up to the Revolutionary War to present ourselves to each other, to the English, and to the world at large as children of the Enlightenment who believe all humans are created equal. Our grievance with the British was that they were not treating us as their equals. As we made this argument, however, we remained faithful to the Puritan legacy that explicitly denied human equality, as evidenced by our treatment of Indigenous peoples and enslaved Blacks. From the contradiction between these two incompatible ideological systems, we've engendered a fractured and paradoxical educational system. We impart to each new generation a perspective on reality that is deeply fissured and unintelligible.

It has been next to impossible for American teachers to nurture an allegiance to truth when a primary duty has been to initiate their young charges into acceptance of the cognitive dissonance at the center of our politics and economic system. We regard it as a duty of teachers to help our students appropriate a fundamental misrepresentation of our principles. It should not then be a surprise if teachers fail to warn students of the damage to habitat that our warming of the planet is now causing—the propagation of convenient misconceptions has *always* been regarded as part of their responsibility. "Allegiance to truth" has never been an educational goal we've thought we could allow ourselves. And it is not just climate change that is being ignored—there is no limit to the amount of fake news that is now being unapologetically shared. For this reason, one might argue that all K–12 education should be free and that later student debts should be forgiven. It seems unfair that one be charged for being told so many incompatible "truths."

Can an Aristotelian perspective be regained to contend against the reigning Puritanism of our time? It's possible, but significantly, many thoughtful people say they do not expect it. Surely making a change will require more from us than we have been willing to give in the recent past.

We know what we must teach:

We must teach the young to not eat meat. Raising livestock requires enormous amounts of water and feed, crowds out forests, and releases harmful methane into Earth's atmosphere; and eating meat itself can heighten disease risk.

We must teach the young to economize on motorized travel, at least as long as it depends on fossil fuels. Cruise ships, for instance, are a luxury that neither the young nor old can afford.

We must teach the young to care for the ocean so that we stop plundering natural fisheries and turning seas into great garbage pits.

But we must teach more than the "thou shalt nots." We must teach critical thinking, gymnastics, the creative arts—those subjects that open the possibility of humans fulfilling their potential and inhabiting a world worth living in.

The generation currently running (and ruining) the planet seems caught in Weber's iron cage—fixated on making as much money as possible, enthralled to demonstrating their Puritan election to grace. They want to own multiple mansions, yachts, and islands as well as other peoples—whose lives seem increasingly for sale. Even politicians who seem "in service to the people" want to retire wealthy.

Our leaders today are arguably the most irresponsible America has ever had. And We the People live in terror of making any sacrifice for our young that they don't really need.

To bring us full circle, I return to the question I began with: What will we tell our grandchildren?

That we thought someone else would solve climate change?

That we knew the solutions but were too lazy to change our behavior?

That we valued profit more than the environment our children would inherit?

That it was too hard, so why bother—and besides, nothing was our fault?

I fear that in fifty or a hundred years our grandchildren will look back on us in disbelief and despair over the lives we have left for them.

Acknowledgments

I would like to express my gratitude to some people who have had a hand in creating this book. Thank you to Karin Wiberg and her team at Clear Sight Books for turning my initial manuscript into a professional finished product; to Liz Grady, who obtained image licenses and provided valuable guidance; to Joe Sottile, who gave feedback on the manuscript from the beginning; to Dr. Raj and her assistant, and Dr. Tom and his nurse, and all of Kaiser, who kept me alive; to Ben Santer, an outstanding climatologist who occasionally sends me his writings; and finally, to my daughter, Natasha O'Neill, for her ongoing editorial feedback, moral support, and love.

Annotated Bibliography

This bibliography may seem haphazard. It addresses works in population growth, agriculture, global warming, meat eating, the education of the young, use of the media, and more. An alert reader will see how one book relates to another. And no doubt some excellent works will go unmentioned. I present works subjectively—those that have influenced me in the development of my manuscript. Some themselves have enlightening annotated bibliographies. (We live in a golden age of expository commentary. A major problem is that the number of people in America willing to read and profit from such commentary is limited and shrinking.)

Asimov, Isaac. *Earth: Our Crowded Spaceship*. John Day, 1974.

This work came out a handful of years after Paul Ehrlich's *The Population Bomb.* Asimov lists Ehrlich's book as something for "further reading," and he would be astounded today by the quaint notion entertained by many academics that Ehrlich had issued a false alarm regarding population growth. The timetable may be off a bit, but Ehrlich's prediction of distress is registered every day now in reports of refugeeism and hunger.

Berry, Thomas. *The Dream of the Earth*. Sierra Club Books, 1988.

Berry's work is not prominently discussed now, a decade and a half after his death, but not because his reflections are no longer relevant. Note, for instance, his sarcasm: "Not only does the corporation provide life's most satisfying experiences, the corporation transforms the entire society in the basic modalities of its functioning. It provides jobs! By having jobs we make money, and with money we buy everything. We no longer grow our own food; we buy our food. We are no longer sustained in the ever-renewing cycle of nature; we live now within the cycle of industrial production and consumption." The deadening effect of the Puritan work ethic as it destroys the taste buds of the soul is referenced here. If Berry is slipping from notice, it's because our popular culture is so far gone that we lack the energy to grasp what he's talking about.

Bourne, Joel K., Jr. *The End of Plenty: The Race to Feed a Crowded World*. W.W. Norton, 2015.

Bourne is a former senior editor at *National Geographic*. His book is a beautifully lucid wake-up call, with the following dedication: "For my mother, who taught me to love language. For my father, who taught me to love the land. And for all those who toil each day to feed the world."

Carroll, James. *Crusade: Chronicles of an Unjust War*. Metropolitan Books/Henry Holt, 2004.

Carroll contributes richly to the theme of my own work. His book is a long howl of protest over our unrepented destruction of Iraq. Toward the end of the book, he writes: "The Manichaeism of the Puritans, the murderous self-righteousness of Indian killers, the Truman Doctrine's assumption of a bipolar division between global forces of light and dark, Ronald Reagan's easy characterization of an 'evil empire'—again and again a simplistic vision imposes itself on a complex world."

Ellsberg, Daniel. *The Doomsday Machine: Confessions of a Nuclear War Planner*. Bloomsbury, 2017.

Ellsberg was a nuclear strategist when he was at RAND Corporation. His book attempts to share the conviction he arrived at through a steady act of reflection over decades: There is no safe use of nuclear weapons. The chapter on the Cuban Missile Crisis is particularly riveting. (It's amazing we're still here to talk about the crisis.) Up until his death in 2023, Ellsberg was becoming increasingly concerned that use of nuclear weapons, even if it didn't quickly end most of our species, would confirm the theory that a "nuclear winter" would come as a consequence and that in the subsequent decade many would die from the nuclear atmosphere, which would block sunlight and close down agriculture. If the book hasn't alarmed everyone, I suspect it's because many are afraid to read it.

Foer, Jonathan Safran. *We Are the Weather: Saving the Planet Begins at Breakfast.* Farrar, Straus and Giroux, 2019.

Foer does an excellent job of presenting the epistemological state in which we find ourselves. As Cardinal Newman said, there's the kind of knowledge—Newman called it "notional knowledge"—that our mind takes hold of and uses to entertain itself; and then there's "real knowledge," which takes hold of us and changes our behavior. On his own behalf, Foer wants to shake off the notional knowledge that keeps him in a state of suspended animation; he wants this so he can begin living authentically, and he wants the same liberation for society in general. Quite rightly, Foer sees everything as frivolous nowadays unless it's favorable to a change of lifestyle. For this reason, he begins with breakfast. He speaks to the reader with an engaging personal voice, and his book is a wonderful reaching out to those who pretend that climate troubles make no claim on their attention—and that's most people.

Freud, Sigmund. *Civilization and Its Discontents.* Doubleday, 1958. [First published in 1930.]

At a certain point in my life, I was deluged by testimonies of academic savants that Freud was passé. It occurred to me that a man who was assigned so many funerals must still be alive. The secret of Freud's survival is that he wasn't a Freudian and that he lived less by the answers he gave than by the questions he asked. He ventured constantly where others feared to go. In *Civilization and Its Discontents*, he raises many issues. He refuses any number of standard conclusions, and his own positive conclusions are offered as tentative. Writing in the aftermath of the First World War, he postulated that we invented civilization and culture because we realized our instinctual life was so chaotic that we'd annihilate our species if we didn't. Civilization was a restraint against aggression. In creating it, however, we had to repress our sexuality. In turn, this act heightened our frustrations. At the end of the essay, he tells us he hopes that Eros may somehow rescue us from the predicament of our species. While Freud makes no mention of Puritanism or Max Weber in this book, the thinking engaged by Max Weber on Puritanism's demoralizing effect on human sensibility can, I think, claim some confirmation from Freud's negative reflections on civilization.

Gelbspan, Ross. *Boiling Point: How Politicians, Big Oil and Coal, Journalists, and Activists Are Fueling the Climate Crisis—and What We Can Do to Avert Disaster*. Basic Books, 2004.

Gelbspan can be ranked among conscientious journalists as the grandfather of notification regarding global warming. He was at Stockholm reporting on the United Nations Conference on the Human Environment in 1972. This marked the beginning of a global sense of responsibility for the state of the planet's environment. In 1997, Gelbspan's *The Heat Is On* was published. (I can see why people avoid letting themselves know about global warming. From the day I encountered this book, I haven't been relaxed about the future of our species. Gelbspan is transparently honest.) Since *The Heat Is On*, Gelbspan has updated the case regarding global warming with *Boiling Point*. The best way to dodge the cogency of the case he makes is immediately to burn the unread book in the fireplace.

Hawken, Paul, ed. *Drawdown: The Most Comprehensive Program Ever Proposed to Reverse Global Warming*. Penguin Books, 2017.

This compilation of measures by which the climate crisis can be mitigated is so dazzling it may produce vertigo. The photography is marvelous, the assembling of expert testimony regarding solutions is rich, and the effort to present mathematically what a particular innovation would achieve toward a reduction of danger and an enhancement of wholesomeness is admirable. What I fear is that the menu of operations may be so awesome as to overwhelm appetite. What one needs to keep in mind when one works with this book is that much of it requires an economic-political federation—a coming together—in order to implement it. I suggest that readers address one section at a time—or, to state the obvious, that different organizations need to commit to different parts of an immense agenda of change in lifestyle. One should have this book at hand as a valuable resource, taking active interest in particular suggestions as opportunity occurs.

Hoggan, James. *Climate Cover-Up: The Crusade to Deny Global Warming*. With Richard Littlemore. Greystone Books, 2009.

We have all been so complicit in the programs whereby we have been deceived that it's unlikely the perpetrators of deceit will ever be called to account. For all of that, they can still be charged with general contempt. This book by Hoggan and collaborator Littlemore can be regarded as a kind of rogues' gallery. It's still important to identify the archvillains, and some are found where you might not expect. Some of them are scientists. Others are executives in our mainstream news. Others have maintained the flow of gasoline without which we could not drive the cars we have designed. There are dignitaries in the ranks of America's Catholic Church and on our Supreme Court who are about as innocent as Iago in Shakespeare's *Othello.* Some pretend to educational prestige while merely running day-care centers. We should go after these people, not to exercise a desire for revenge, but because we need to disempower them before they've ruined everything. Hoggan and Littlemore may help us do this.

Klein, Naomi. *On Fire: The (Burning) Case for a Green New Deal.* Simon & Shuster, 2019.

One can savor here Naomi Klein's eloquence and thoughtful reflection in her account of the *Deepwater Horizon* catastrophe. In it, she invites us to see that the same oil producers and geologists who have mendaciously denied that our species is causing global warming have also pretended to know what they're doing in their offshore drilling for oil. That her words have not provoked outrage is not her fault. The news is just too sad for most of us to absorb.

Mander, Jerry. *In the Absence of the Sacred: The Failure of Technology and the Survival of the Indian Nations.* Sierra Club Books, 1991.

Mander's book develops a theme that Naomi Klein mentions. Klein says that with the 1600s, we members of Western civilization began thinking of the earth as a machine that we could manipulate to our purposes. Mander regrets that we've lost our sense of the sacredness and organicity of the earth around us. He's right to do so. If we can't rediscover the magical mystery—the inexplicable wonder and beauty—of the earth, we're going to lose it in the foreseeable future as the welcoming habitat it has been.

Marcuse, Herbert. *Eros and Civilization: A Philosophical Inquiry into Freud.* Vintage Books, 1962. [First published in 1955 by Beacon Press.]

Some have criticized this work by Marcuse as too optimistic. Written at the start of the 1960s, it does have a kind of optimism that seems out of reach today. Yet it remains relevant. If I have understood him, Marcuse is suggesting the need for a kind of civilization that is not as discontent-prone as the civilization Freud discusses in *Civilization and Its Discontents.* Citing various hints within Freudianism, Marcuse suggests that civilization need not be so repressive of sexual instinct as ours is. Marcuse does mention Puritan era and Victorian era repression. What happens when one promotes a Puritanical (that is, a Calvinist) approach to sexuality is not that there's less sexuality but rather that sexuality is less wholesome. In its guilt-ridden form, it is less likely to function as a springboard to eros—and to the love and cherishing of humanity in general and to the rich vistas and opportunities that beckon human beings.

McKibben, Bill. *Falter: Has the Human Game Begun to Play Itself Out?* Henry Holt, 2019.

In 1989, McKibben's first book, *The End of Nature*, was published, and he has been a wonderful voice for environmentalism ever since. In his more recent book, *Falter*, he indicates he still has hope for the future. What's unique is McKibben's gift for anecdotes. He finds or invents interesting ones and recites them well. They serve as guideposts for action by those who remain hopeful as global warming progresses.

Miller, Perry. *Errand into the Wilderness.* Belknap/Harvard University Press, 1956.

This work contains an essay titled "The Marrow of Puritan Divinity." In that essay Miller says, "In dogmatic Calvinism morality could exist only as a series of divine commands. It has no other basis, and to Calvin it needed no other. The covenant theology is a recognition on the part of a subsequent generation that this basis was inadequate, that it reduced morality to an arbitrary fiat, that it presented no inducement to men other than the whip and lash of an angry God." Miller shows how the revision of Calvinism by promoting the theme of a covenant with God (with some free will after all!) improved Calvinism's prospects for becoming a viable religion in Massachusetts and, eventually, for the United States in general. This softening meant that the doctrine of the Elect could then continue and have (in every moment since) a great effect on our domestic and foreign policies.

Oreskes, Naomi, and Erik M. Conway. *Merchants of Doubt: How a Handful of Scientists Obscured the Truth on Issues from Tobacco Smoke to Global Warming*. Bloomsbury, 2010.

One of the distressing facts about this book is that it is a labor of great specificity, and yet the practices of deflection on which it turns a spotlight continue. One reason: In today's world of social media, one can vividly imagine The Truth—or at least "the truth for you." Another reason: The very virtues of this book may work against it. The conscientious precision with which all the speakers, all the arguments, and all the responses to arguments are presented may overwhelm the reader. To not be overwhelmed, I suggest the tactic of selecting only one theme to address on any given day and letting the rest of the book wait. (This book also contains a magnificent commentary on the heroic life of Ben Santer. It seems his enemies would have been happy to drive him to suicide but haven't been able to do it. His fortitude prevails.)

Perkins, John. *The Secret History of the American Empire: Economic Hit Men, Jackals, and the Truth About Global Corruption.* Dutton, 2007.

Were I a relative or friend of Perkins, I might have advised him against publishing this book. I would have said, "John, these malefactors of wealth and power are not going to let you live peacefully if this book comes out." I'd have been too risk averse to publish it myself. Perkins is still up and about, some two decades after publication. This book documents his indictment of American exceptionalism by providing names and dates and accusations that indicate how cruel we've been to people in other countries—much like the Puritan slaughter of the Pequot in 1637—over the past two centuries.

Rich, Nathaniel. *Losing Earth: A Recent History*. Farrar, Straus and Giroux, 2019.

Here is a poignant account, leading particularly to 1989, when the senior Bush presidency threw in the towel on taking global warming seriously. In the subsequent eight years, Clinton was busy taking Republican policies and preempting them. While he did try for peace in the Middle East and had some success in reducing the sale of semi-automatic weapons to civilians, he seems to have muzzled Al Gore, whose most notable assignment was debating Ross Perot on free trade. Regarding Rich's book, the most eloquent section is the nearly thirty-page afterword. Read it and weep.

Ricks, Thomas E. *Fiasco: The American Military Adventure in Iraq*. Penguin, 2006.

Ricks must have been closely connected to the military to have been able to write the book he did. He does not villainize the soldiers; they had the impossible task of making the "adventure in Iraq" rational when it had no well-defined goals. The invasion and smashing of Iraq is arguably the most criminally irresponsible foreign policy our country has ever perpetrated. The tone of Ricks's account is not a howl of outrage; it is a quiet, carefully sourced account of who was doing what for what discernible reasons.

Segal, Charles M., and David C. Stineback. *Puritans, Indians, and Manifest Destiny*. Putnam, 1977.

That this book has been generally neglected shows how far gone we are in substituting hagiography for genuine history. The book connects key points: the Puritan founding of Massachusetts Bay Colony, the genocide that the Puritan Elect carry out against the Pequot, the convenient "discovery" that it's manifestly our God-given destiny to control most of North America. These points are important because, while the book doesn't explicitly carry them further, they establish a context for our subsequent unilateral declaration that the Monroe Doctrine justifies our assumption of hegemonic control of the entire Western Hemisphere. It is in this context that we continue on after the Second World War by putting ourselves forward as "the leader of the free world." God help those who don't want to be freed in this way!

Smith, Adam. *The Wealth of Nations, Books I–III*. Penguin Classics, 1982. [First published in 1776.]

This edition does not include the famous sentence on the "invisible hand"; that phrase occurs in Book IV, Chapter II. One could wish it didn't occur at all, because it seems to have been taken by many as the sovereign key to Smith's economics—which it is not. Smith was writing here about the businessman's freedom to work from his self-interest. It's important to recognize that this "self-interest" is not the selfishness of, say, Ayn Rand. Smith is saying the individual businessman [businessperson] is likely to know better than a bureaucrat how to produce a good product or a good service. Smith was never an advocate of a cutthroat, scorched-earth corporate capitalism. He writes: "People of the same trade seldom meet together, even for merriment and diversion, but the conversation ends in a conspiracy against the public, or in some contrivance to raise prices." A bit later, he writes, "The pretence that corporations are necessary for the better government of the trade, is without any foundation." As Smith opposed mercantilism, so he opposed monopolies in general, writing: "Monopoly . . . is a great enemy to good management." One should not let his kidnappers speak for him; Smith has a sweet voice. He believed in change and improvement—hence, his book—but, like his Anglo-Irish contemporary, Edmund Burke, he favored incrementalism: Venture some change; look at it; make the next change in the light of the last. (This is precisely what we should have been doing regarding the threat of global warming from the 1950s on.)

Thoreau, Henry David. *Walden and Civil Disobedience: Complete Texts with Introduction, Historical Contexts, Critical Essays.* Edited by Paul Lauter. Houghton Mifflin, 2000. [*Walden* was first published in 1854 by Ticknor and Fields, a predecessor to Houghton Mifflin; the essay that came to be called "Civil Disobedience" was first published in 1849.]

Walden lines up with the more recent works by Jerry Mander and Thomas Berry that I've cited. It enables the reader to understand that the crisis we are in is not primarily a chemical crisis or a technological crisis or a political crisis; ours is rather a crisis of the spirit. In America now we find ourselves in a vicious cycle: That which we do not find precious, we do not take care of; that which we do not take care of, we do not find precious. Thoreau says a person can find out more about reality by studying what's between his front door and his front gate than he can by traveling around the world. As for the essay on civil disobedience, it's more relevant than ever. Our terrible squandering of opportunities since the end of the Second World War, especially our war of horrors against Vietnam and our invasion and smashing of Iraq, means that we must be alert to the lies and reckless spending of our government—and we must resist.

Wallace-Wells, David. *The Uninhabitable Earth: Life After Warming.* Tim Duggan Books, 2019.

The sorrow this book can induce is beyond description. Wallace-Wells does not generate the sorrow; he merely takes away our blinders.

Weber, Max. *The Protestant Ethic and the Spirit of Capitalism.* Translated by Talcott Parsons. Charles Scribner's Sons, 1958. [First published in Germany in 1905.]

This book provides us an opportunity to know ourselves as few books do. When I've attempted to call Weber to the attention of colleagues, I've been amazed how many of them "know all about him" but clearly have never read him nor have any idea what he says. One told me she was quite sure Weber said nothing about aesthetics because "he wasn't a psychologist." Another told me Weber was "kind of a Marxist-lite." A third, a professor, no sooner heard me mention Weber than he brushed him to one side, saying, "Well, Talcott Parsons has come along since." (Parsons was Weber's translator and influential interpreter in American academia, not a figure who renders Weber obsolete.) To understand American exceptionalism, to understand American militancy, to understand American selfishness and scorched-earth capitalism, one can hardly do better than to read Weber.

Williams, William Appleman. *The Tragedy of American Diplomacy.* Dell, 1962. [This is a revision of a 1959 publication by World Publishing Company.]

I'm afraid almost no one reads this fairly short and elegantly argued analysis anymore. That's too bad. Williams comes across as a scholar who spent hours and hours of his life pondering the question: "What are we Americans up to?" He takes on US Secretary of State John Hay's Open Door policy, written in a series of notes from 1899 to 1900, and its subsequent history. He sees why Japan felt crowded. He sees, after the Second World War, why Stalin said no to the Marshall Plan. He can greatly help one to come to an understanding of how we squandered our unique situation as the world's only superpower; the consequences have been considerable and an understanding of what we did may be a fertile ground for undoing it.

Wu, Tim. *The Attention Merchants: The Epic Scramble to Get Inside Our Heads.* Vintage/Penguin Random House, 2017.

The writing is temperate, but the continuing damage to our prospects for a thoughtful life is unnerving. If you wonder why our lives these days seem less wholesome than they used to be, Wu is ready with a detailed account of how we got here and why we shouldn't stay here.

Zinn, Howard. *A People's History of the United States.* Harper Perennial, 2005. [First published in 1980.]

The Parson Weems school of American history is displeased with this book. History, they say, should make you feel good, but many parts of this book are likely to make a reader anything but glad. Consider that much we are taught in school never happened, and much that did happen we have not been taught about. In reading Zinn, you may experience one eureka moment after another. This can be exhilarating, but also, as George Santayana said, "Those who cannot remember the past are condemned to repeat it." Knowing the mistakes we've made is a powerful way toward liberation.

Index

Abbott, Greg, 118
ABC news network, 28, 100, 107, 109
Adams, John Quincy, 13, 68
addiction, 105
advertising, 99
Afghanistan, 74
Africa, 15–20, 51, 77
agnosticism, 53, 123
agriculture, farming, 11, 13, 23, 32
Alexander VI (Pope), 37
America. *see also* Indigenous Americans
  addiction, suicide, 105
  American Dream, 14
  American Puritanism, 44, 55, 83
  colonialism, 11, 16, 51, 52, 61–64
  conflicting traditions, 71, 104, 122
  Constitution, 75
  Declaration of Independence, 65, 75
  democracy, 73, 75
  Federalists, 68, 69
  founding fathers, 52
  justifies anything, 39, 65, 74, 100
  land seizures, 13, 43, 48, 51, 68, 80
  self-aggrandizement, 45, 75, 122
  slavery, 51, 54, 68, 75–81, 124
  sphere of influence, 74
  submit to or die, 64, 81
  superconsumers, 46
  US war department, 14
  USDA, 29
Anglicanism, 43, 56
Aquinas, Thomas, 116, 120
Aristotle, 77, 96, 107, 112, 115–21, 119
Arrhenius, Svante, 31
Assange, Julian, 101
atheism, 53, 123
Athens, Greece, 119
The Attention Merchants (Wu), 100, 108
autonomy, 40, 45, 118
Bangladesh, 21–25
Barnouw, Erik, 109
Basquiat, Jean Michel, 5
Bay Colonists, 52, 61–64
Bay of Bengal, 23
Beard, Charles, 80

Beard, Mary Ritter, 80
Belgium, 16, 18
Bezos, Jeff, 14
Bible, 40, 56, 66, *see also* religion
Bikini Atoll, 102
bin Laden, Osama, 74
Bonhoeffer, Dietrich, 57
Borlaug, Norman, 29
Bourne, Joel K., Jr., 27, 28–29
BP (formerly British Petroleum), 94
Brahmaputra River, 25
Britain, 11, 21, 24, 53, 124
broadcast media, 101, *see also* news stations
Buddha, 112
Burke, Edmund, 7
Bush, George W., 74, 100
businesses
- buy lawmakers, 91, 98, 102
- capitalism eternal, 45
- employees, workforce, 84–87
- indifference of, 90, 93
- mercantilism, 89, 92
- monopolies, 92
- profits, 83–98
- social consciousness, 84, 85, 94

Calvinism
- chosen, predestined, 12, 57, 66, 75
- origins, 37–41
- Pilgrims, 51

Canaanites, 56, 66
capitalism
- ethics and, 44, 45, 86
- force-feeding to Europe, 69
- free-market economics, 86
- "iron cage", 45, 97, 109, 126
- profit at any cost, 84–85, 93
- Puritan roots, 41, 44
- pursuit of wealth, 78–79, 81, 107, 112
- scorched-earth, 65, 83, 101

carbon dioxide, 12, 92
carbon footprints, 4
Catholicism, 17, 37, 46, 55, 120
CBS news network, 28, 100, 105, 109
Charles I (King of England), 56
Charles II (King of England), 76
Cherokee people, 68
Christianity, 43, 55, 96
Churchill, Winston, 69
Citizens United, 91, 102
civil wars, 19, 73, *see also* wars, incidents
Clay, Henry, 13
climate change
- carbon dioxide, gases, 12, 31, 92, 106
- climatology, 32
- conservation, 26, 30
- food supply, 13, 23, 28, 32
- global warming, 23, 31–34, 92, 103, 125
- natural disasters, 28, 104, 106
- population growth, 4, 28
- predictions, projections, 28
- water supplies, 23, 27, 29

cognitive dissonance, 104, 124
colonialism, 11, 16, 51, 52, 61–64
common good, 79–80, 92, 96, 115, 119, 120–23, 124
communism, 64
Confucius, 116, 118, 121

Connecticut Colony, 63
conservation, 26, 30
conservatism, 83, 93, 95
Constitution, 75
corporations. *see* businesses
Counter-Reformation, 38
creation, 58, 65
Cuba, 49, 74
Custer, George Armstrong, 13
Darwin, Charles, 119
Declaration of Independence, 65, 75
deism, 53
Delian League, 119
democracy, 22, 73, 75
Democratic Party, 46, 68, 80, 98
DeSantis, Ron, 118
destiny, 6, 38, 47, 49, 97, 124, *see also* predestination
DuPont, 101
Dust Bowl, 13
Dyson, Freeman, 107
economics
  capitalism, 86
  cognitive dissonance, 124
  ethics endorses, 45, 86
  scorched-earth approach to, 83
  secular Puritanism, 71–80
education, 93, 95, 116–26
Elect class, 47, 54, 63–66, 106, 124
Ellsberg, Daniel, 101
energy boom, 11
England, 10, 43, 55–56, 76
Enlightenment
  European origins, 53–55
  federalists, 68, 69
  origins, 6–7
  perspective, 71
  quandary of, 65, 74, 124
Enron, 92
equality, 54, 75–78, 117, 122, *see also* slavery
ethics
  animal habitats, 4
  Aristotle, 115
  capitalism, 44, 45, 86
  discipline vs., 85
  economics, 45, 86
evolution, 119
farming, agriculture, 11, 13, 23, 32
Federalists, 68, 69
Fermi, Enrico, 112
First Amendment, 110
Food and Agricultural Organization, 28
food supply, 9, 13, 23, 28, 32, 125
Fourier, Joseph, 31
Fox News, 86
France, 7, 53, 54, 55, 71
Franklin, Benjamin, 45
freedom, 39, 40, 58, 66, 92, 118, *see also* slavery
free-enterprise system, 86
French Revolution, 7, 55, 71
Friedman, Milton, 44, 45, 79, 83–98, 110
Gandhi, 21
Ganges River, 25
genocide, 13, 15, 19, 21, 61
George (King of England), 77
Germany, 16, 53
Girard, René, 111
global warming. *see* climate change

God, 38–40, 44, 45, 53, 56–59, 63, *see also* religion
government. *see also* politics
  citizens and civic duty, 80
  Citizens United, 91, 102
  common good, 79–80, 92, 96, 115, 119, 124
  democracy, 22, 73, 75
  for sale, 91, 98, 102
  function of, 79
  "invisible hand", 14, 92
  involvement, intrusion, 80, 83
  optional, 78
  responsibilities, 79, 92
Great Depression, 69
Green Revolution, 29
greenhouse gases, 31–32, 106, *see also* climate change
Guantanamo Bay, 74
Gulf of Mexico, 94
Gulf of Tonkin, 74
habitats
  humans configure, 4, 5
  pollution and, 14, 84, 90, 93
  population affects, 12, 14, 35
  students unaware of threats to, 124
Halberstam, David, 108
Hamilton, Alexander, 7, 68
Hansen, James, 30, 34, 103
Harding, Warren, 69
healthcare, 16, 73, 79
Hegel, G. W. F., 7–8
Henry VIII (King of England), 43
The Hidden Persuaders (Packard), 100
Hobbes, Thomas, 80
Homan, Jim, 64
human rights, needs, 75, 78, 79, 80, 81, 111, *see also* equality
Humboldt, Alexander von, 31
Hume, David, 123
Hutu people, 15–19
idealism, 48
India, 21
Indigenous Americans
  Cherokee, 68
  land "unoccupied," seized, 13, 43, 48, 51, 68, 80
  Massachusetts tribe, 43
  Mystic Fort massacre, 62
  Narragansett, 62
  Niantic, 62
  Pequot, 61
  Sioux, 13
  told 'submit or die', 80
  Trail of Tears, 68
Industrial Revolution, 120
infant mortality, 16, 22
Institute for Demographic Studies (INED), 23
Intergovernmental Panel on Climate Change (IPCC), 27, 103
International Geophysical Year (IGY) initiative, 32
"invisible hand", 14, 92
Iraq, 70, 74
"iron cage", 45, 109, 126
irrigation, 13, 23, 29–30, 32
Jackson, Andrew, 13, 66, 67, 75
Japan, 71
Jefferson, Thomas, 51, 54, 66, 67, 75, 77
Jesus, 40, 58, 112, 116, *see also* religion

Jim Crowism, 68
Jobs, Steve, 14, 111
Johnson, Lyndon B., 74
journalism, 99–110
Judah, Theodore, 88
Keeling Curve, 32
Kennedy, John F., 74
Keynes, John Maynard, 89
King, Martin Luther, Jr., 57
Koch, Charles, 14, 107
land
  Indigenous peoples', 14, 43, 48, 51, 68, 80
  ownership formula, 75
Lawrence Livermore National Laboratory (LLNL), 3
Leo XIV (Pope), 120
liberalism, 70, 80, 83, 85, 98, 117–18
Lincoln, Abraham, 118
Lippmann, Walter, 108
livestock, 29, 125
Locke, John, 48, 51, 75–81
Louisiana Purchase, 80
Luther, Martin, 37
Machiavelli, 121
Madison, James, 51, 66, 75
Magnus, Albertus, 120
Malthus, Thomas Robert, 9–13, 17
Mander, Jerry, 108
Manifest Destiny, 47
Manning, Chelsea, 101
Marcuse, Herbert, 108, 111
Marshall, John, 68
Marshallese people, 102
Mason, John, 63
Massachusetts Bay Colony, 52, 56, 61, 63
Mayflower, 52
McLuhan, Marshall, 108
media, 101, 102–13
mercantilism, 89, 92
Mexico, 49, 94
Miller, Perry, 48, 55
Milton Friedman, 121
money. *see also* capitalism
  function of, 89, 96
  materialism, 86
  prosperity gospel, 45, 106, 110
  pursuing wealth, 78–79, 81, 107, 112
Monroe Doctrine, 49, 74
Monroe, James, 51, 66, 75
Monsanto, 101
Moonves, Les, 106, 109
moral agency, 39, 40, 48, 58
Muir, David, 102, 110
Murrow, Edward R., 100, 108
Musk, Elon, 14, 122
Myanmar, 23
Mystic Fort massacre, 61
Nagasaki, Japan, 72
narcissism, 86, 112, 116
Narragansett people, 62
National Geographic, 27
national security, 79
natural disasters, 28, 104, 106
NBC news network, 28, 105
neo-Malthusianism, 12
New Deal, 70
New England, 43, 48, 55, 75
New Testament, 40, 56, *see also* religion

New York Times Magazine, 83
news stations, 28, 86, 94, 100, 107, 109
Niantic people, 62
Nicomachean Ethics (Aristotle), 115
Notes on the State of Virginia (Jefferson), 67
nuclear power, 3, 30, 49, 65, 72
Nuremberg trials, 76
Obama, Barack, 74
oceans, 25, 28, 35, 125
Oldham, John, 62
optimism, 6–8, 122
overpopulation, 6, 15, *see also* population
Packard, Vance, 100, 108
Pakistan, 21, 74
Paley, William, 100, 105
parenting, 116
Partition, 21
patriotism, 118
Paul (Apostle), 38, 56
PBS News Hour, 94
Pelley, Scott, 28, 109
Pequot people, 61
Pericles, 119
pessimism, 9–14
Pilgrims, 51, 62, *see also* religion
Plato, 115–17, 120
Politics (Aristotle), 77, 117, 119, 121
politics
- Bangladesh, 21
- cognitive dissonance, 124
- communism, 64
- conservatism, 83, 93, 95
- Democratic Party, 46, 68, 80, 98
- Enlightenment vs. Puritanism, 64–70
- liberalism, 70, 80, 83, 85, 98, 117–18
- politicians for sale, 91, 98, 102
- religion and, 54
- Republican Party, 46, 68, 81, 98
- socialism, 84, 90
- WWII, 71, 76

Polk, James, 75
pollution, 14, 84, 90, 93
Pope Leo XIII, 120
population
- Bangladesh, 22
- carbon footprints, 4
- colonialism, 11, 62
- food production vs., 9–11
- global warming, 4, 28
- habitat maintenance and, 12
- infants, 16, 18, 22
- medical care, 16
- misery response and, 19, 25
- overpopulation, 6, 15
- predictions, projections, 9, 17, 23, 28
- Rwanda, 15–20
- succeed into failure, 10, 15

positivists, 122
Pouillet, Claude, 31
poverty, 7, 10
predestination, 38–40, 44, 54, 56, 63
The Protestant Ethic and the Spirit of Capitalism (Weber), 41
profits, 83–98, *see also* businesses
property, 51, 75, 78–79, 80, 86
prosperity gospel, 45, 106, 110

Protestantism, 17
Puritanism
  Elect, entitlement, 39, 63–66, 75, 124
  Pequot massacre, 61–70
  roots, ideology of, 43–49, 51–58, 86, 97
  secular economics of, 71–80, 96
Reformation, 37
refugees, 23, 26
Reign of Terror, 7, 55, 71
religion. *see also* Puritanism
  agnosticism, atheism, 53, 123
  Calvinism, 12, 37–41, 51, 57, 66, 75
  capitalism, common good, 46, 120
  Catholicism, 17, 37, 46, 55, 120
  Christianity, 43, 55, 96
  deism, theism, 53
  God, 38–40, 44, 45, 53, 56–59, 63
  Jesus, 40, 58, 112, 116
  New Testament, 40, 56
  Pilgrims, 51, 62
  politics and, 54
  predestination, 38–40, 44, 54, 56, 63
  prosperity gospel, 45, 106, 110
  Protestantism, 17
  Reformation, 37–41
  sin, 56, 58
  "vessels of election, wrath", 56
Renaissance, 37
Republican Party, 46, 68, 81, 98
Rerum Novarum (Leo XIII), 120
Revolutionary War, 13, 65, 75, 124
Rohingya people, 23
Roman Catholics, 17, 37
Roosevelt, Franklin Delano, 69
Roosevelt, Theodore, 48
Rosenberg, Julius & Ethel, 86
Roundup, 101
Royal African Company, 77
Rwanda, Africa, 15–20
Santer, Benjamin, 103
Sarnoff, David, 105
Sartre, Jean-Paul, 66
sexism, 77, 117, *see also* equality
Shell Oil, 92
sin, 56, 58
Sioux people, 13
slavery, 51, 54, 68, 75–81, 124
Smith, Adam, 7, 14, 83, 89, 92
Snowden, Edward, 101
"The Social Responsibility of Business Is to Increase Its Profits" (Friedman), 91
socialism, 84, 90
society
  ambitions of, 12, 86, 90
  educating, 95, 116–26
  relationships, 78
  socially conscience businesses, 84, 85, 94
Socrates, 117
Soviet Union, 69, 74
Spanish-American War, 49
Stone, Christopher, 88
Stone, John, 62
Stuart kings, 43, 55
superconsumers, 46

technology, 3, 11, 122
Teflon, 101
theism, 53
Third-World countries, 12
Thoreau, Henry David, 115
Tillerson, Rex, 14
Trail of Tears, 68
tribal rivalries, 15, 18
Truman, Harry, 69
Trump, Donald J., 46, 86
Turner, Frederick Jackson, 13
Tutsi people, 15–19
US Department of Agriculture (USDA), 29
US Department of War, 14
Van Buren, Martin, 68
Vietnam, 64, 74
wars, incidents
  Bikini Atoll, 102
  captives of, 76
  civil wars, 19, 73
  Cuban Missile Crisis, 74
  French Revolution, 7, 55, 71
  Mystic Fort massacre, 61
  necessary?, 8
  related deaths, 64
  Revolutionary War, 13, 65, 75, 124
  Spanish-American War, 49
  US Department of War, 13
  Vietnam, 64, 74
  World War II, 71, 76
Washington, George, 52–53
water supply
  access and quality, 22, 27, 29
  Bangladesh, 22
  conservation, 26, 30
  desalination, 30
  irrigation demand, 13, 23, 29–30, 32
  ocean contaminates, 125
  ocean level, 25, 35
  predictions, projections, 23, 29–30
  USDA water management report, 29
  vegetarianism and, 125
weapons, 3, 49, 65, 72, 73
Weber, Max, 39–41, 44, 45, 66, 97, 109, 110, 126
Weems, Parson, 52
Whitehead, Alfred North, 57
wildfires, 27, 110
Wilson, E. O., 15, 16, 19
Wilson, Woodrow, 69
The Winning of the West (Roosevelt), 48
Winthrop, John, 43, 47, 52, 55, 65
World Meteorological Organization, 33
World War II, 71, 76
Wu, Tim, 100, 108
Yeats, William Butler, 81
Zinn, Howard, 80
Zuckerberg, Mark, 14

About the Author

TOM O'NEILL has spent a lifetime exploring ethics, history, and human behavior. He trained with the Jesuits, worked as a social worker in Compton, California, and taught in Micronesia, where he learned far more than he imparted. He then spent most of his career teaching philosophy and religion in college classrooms. O'Neill is the author of *Why the Center Can't Hold: A Diagnosis of Puritanized America* (punctum books, 2016) and lives in Livermore, California, where he continues to write on the cultural and moral forces shaping our world.

www.ingramcontent.com/pod-product-compliance
Lightning Source LLC
LaVergne TN
LVHW022048200926
848481LV00003B/834